The Sheep as an Experimental Animal

The Sheep as an Experimental Animal

by

J. F. Hecker
Department of Physiology,
The University of New England,
Armidale, NSW, Australia

1983

ACADEMIC PRESS
A Subsidiary of Harcourt Brace Jovanovich, Publishers
LONDON NEW YORK
PARIS SAN DIEGO SAN FRANCISCO SÃO PAULO
SYDNEY TOKYO TORONTO

ACADEMIC PRESS INC. (LONDON) LTD.
24/28 Oval Road,
London NW1

United States Edition published by
ACADEMIC PRESS INC.
111 Fifth Avenue
New York, New York 10003

British Library Cataloguing in Publication Data

Hecker, J. F.
The sheep as an experimental animal
1. Sheep 2. Laboratory animals
I. Title
636.3 SF375
ISBN 0-12-336050-1

Photoset by Paston Press, Norwich
Printed by Whitstable Litho Ltd., Whitstable, Kent

Preface

Every book is written for a potential group of readers. This book is intended to be of use to those who may, who intend to, or do use sheep as laboratory animals. There are a considerable number of laboratory animals used for experimental research and many have had books written on their use in the laboratory. At first glance, the sheep is not normally considered a laboratory animal. However it is used widely and there are sufficient surgical techniques developed for a book of over 300 pages to have been written on experimental sheep surgery. Apart from one short introductory chapter in that book and a few short articles in journals, very little of substance has been published on sheep as experimental animals. The aim of this book is to fill the gap.

There is a problem in writing a book such as this in the selection of the material to be included. Research for it has produced a large number of items of information relating to sheep as laboratory animals. The publishers have made recommendations as to length and, in order to keep to this length, it has been necessary to be selective and to make decisions as to what should be included. To do this, I have tried to put myself into the position of someone who is not familiar with using sheep for experimentation and then have attempted to cover questions which this person might raise. This creates a problem in that there are many different techniques used in different laboratories for the management of experimental sheep. The methods described have been gleaned from my observations in many laboratories in three continents but they undoubtedly reflect my greater familiarity with laboratories in Australia and with the University of New England in particular. I have also tried to detail items of information that the above person might expect to find in a book of this type. This has also involved much selection on my behalf as I have found that the scientific literature is very large. While I have tried to keep this selection broad, I realize that there are a few places where my interests become apparent.

It has been necessary to omit certain topics which are relevant to sheep as experimental animals. As a result, one area which has been largely ignored is that of growth and development. Also there is little on foetal sheep although there is an immense literature on this topic. Perhaps someone will write a book on the foetus as an experimental animal in which the sheep foetus would undoubtedly be prominent.

I would like to thank those people who have contributed ideas and criticisms to this book. Also I would like to thank the following for permission to reproduce copyright material: Robert E. Habel, Baillière Tindall, Saunders, Paul Parey, Hutchinson and the publishers of the journals *Endocrinology*, *Journal of Endocrinology*, *Behaviour* and *Journal of Physiology*.

August 1982 *J. F. Hecker*

Contents

Glossary

Broken mouth:	Lacking one or more incisor teeth.
Cast:	(a) a sheep that is on its back and cannot right itself. (b) cull.
Cull:	An animal sold as not fit for future production.
Ewe:	Entire female sheep.
Gimmer:	Young ewe (British).
Hogg, hoggett:	A female sheep older than a lamb but not having two permanent teeth.
Lamb:	A young sheep, usually up to the age of about 6 months. Sometimes in American medical journals, the term is used to indicate a sheep that is not mature.
Polled:	Lacking horns.
Ram:	Entire male sheep.
Struck:	Suffering from fly strike.
Wether:	Castrated male sheep.

Chapter 1

Introduction

1.1 The Sheep as a Laboratory Animal

Biological research relies extensively on experimental animals. In the past, much biological research has been done with the traditional laboratory animals such as the mouse, rat, rabbit and guinea pig. While these animals are still usually preferred for the reasons of cost and familiarity, their small sizes make them unsuited for some types of experimentation and larger laboratory animals are required. For these animals, research workers have frequently turned to the dog as there is often a ready supply of discarded household pets, or to farm animals of which there is also a ready supply. Of the common farm animals, the horse and the cow are sometimes used as laboratory animals (for antibody production and occasionally for foetal research) but their high cost of purchase and maintenance and the large space that they require limit their usefulness. Preferred are the dog, pig, sheep and goat.

The dog is the traditional large laboratory animal. Two reasons for this are the ready supply of mongrel dogs in most cities which would otherwise be euthanazed and the familiarity of research workers with the care of dogs as most will have had experience with pet dogs. However the idea of using cast off pets for research is repugnant to many people and especially to members of the antivivisectionist groups. Also in some countries, mongrel dogs can be expensive ($50 is a current price in some cities in the United States) although not nearly as expensive as pure bred Beagles which have been raised especially for research.

The pig is an alternative large laboratory animal and two books, *The Pig as an Experimental Animal*[31] and *The Pig in Biomedical Research*[7] have been published. Pigs have many similarities to man, particularly in their digestive system and skin, but they can grow to an unmanageable size and can be most unpleasant and uncooperative animals in the laboratory. These attributes restrict their use. Goats are also used in some regions as a laboratory animal but in many other regions their supply can be a problem. In nearly all respects, the goat is similar to the sheep and most of the remarks in this book about sheep are applicable also to goats.

This leaves the sheep (Table 1). Sheep are raised in nearly every country and in large numbers in most (Table 2). To a person with a veterinary or agricultural background, there are no great problems with the management of sheep in the laboratory. However to someone with a medical or other non-agricultural scientific training, the sheep initially appears to be a very alien animal. But when such a person starts to use sheep in the laboratory, they find that the potential problems were largely imaginary and that the sheep is actually a very easy animal to manage in the laboratory and to use for experimentation.

The sheep's place in research, albeit comparatively small, goes back a long way. In 1667, Jean Dennis, physician to Louis XIV of France transfused a boy aged 15 years with blood taken from the carotid artery of a lamb. The boy felt "a great heat along his arm"; probably the first description of a transfusion incompatibility.[14] The first

Table 1 A classification of the domestic sheep (after Ryder and Stephenson [35]).

Class:	*Mammalia* (warm blooded animals which suckle their young).
Order:	*Artiodactyla* (even toed or cloven hooved animals (ungulates)—this order includes pigs, hippotamuses and camels).
Suborder:	*Pecora* (have hollow horns which grow continuously—other families are deer (*Cervidae*) and giraffes (*Giraffidae*)).
Family:	*Bovidae* (cattle family—hollow horns which grow continuously).
Tribe:	*Caprini* (goat-like).
Genus:	*Ovis* (other genera are *Capra* (goats and ibex), *Hemitragus* (tahr), *Ammotragus* (Barbary sheep) and *Pseudois* (blue sheep).
Species:	*Aries* (other species are *O. canadensis* (American bighorn), *O. ammon* (Argali) and *O. musimon* (mouflon)).

Table 2 Numbers of sheep in some countries (from [25]).

Country	Sheep numbers
USA	17 724 000
South America	5 772 000
Africa	136 550 000
China	72 000 000
Asia other than China	194 790 000
Great Britain	28 089 000
France	10 218 000
Spain	17 191 000
Europe—Total	121 863 000
USSR	139 086 000
New Zealand	69 722 000
Australia	140 109 000
World total	1 018 300 000

recorded transfusion experiments in England, demonstrated to members of the Royal Society also in 1667, again used sheep blood.[18] In Paris in 1790, a man whose name became a symbol of the French revolution, Dr Guillotin, perfected his "philanthropic decapitating machine" on sheep.[29] And in 1863 in Glasgow 5 years before he first published his findings on antiseptic surgery, Joseph Lister cannulated the jugular vein of a sheep with a loop of Indian rubber tubing and recorded that the blood did not coagulate and obstruct the lumen but formed a lining membrane.[14] This was the first recorded account of the problem of thrombus formation on intravascular devices. Since these humble beginnings, the sheep has become widely used for research and its use continues to increase in parallel with the general increase in scientific publication (Table 3).

Use of the sheep for research falls into three broad areas. The sheep has great importance as a meat animal and as the only wool producing animal and hence considerable effort is devoted to "agricultural" type research which is directed mainly towards improving production of these commodities. This type of research includes study of most sheep diseases and parasites, nutrition and feeding, genetics and breeding, reproduction and growth, and wool characteristics. Most of the agricultural type of research is well established and well known and hence less emphasis will be given to it in this book.

Table 3 Numbers of citations for laboratory animals in biological abstracts.

Year	1970	1975	1980
Cat	1 895	2 462	2 090
Dog	4 191	3 346	3 904
Goat	350	373	457
Guinea pig	1 981	2 174	2 193
Mouse	5 350	8 238	11 785
Pig	1 282	1 484	1 652
Rabbit	4 598	5 359	5 764
Rat	10 891	15 033	19 499
Sheep	1 806	2 600	2 782

Numbers include citations under related terms (e.g. mouse, mice, murine).

A second type of research is medical research done with the aim of obtaining a better understanding of human disease and its treatment. At first consideration, the sheep is a surprising choice for such research but it does have some important advantages and in many aspects it is possibly more similar to man than some of the traditional laboratory animals. Several examples of the use of the sheep for medical type of research are given later. Often the reason for its selection for medical research lies in the similarity in size of the sheep to man and to its suitability for chronic experimentation after surgical modification.[19] Possibly only for the dog does there exist a wider range of experimental surgical techniques.

The third broad research area is that of general biological knowledge. The sheep is a convenient example of the group of herbivorous animals which have ruminant type digestion and this deserves and receives study in its own right. The sheep's importance however is wider than this and includes aspects which are either peculiar to the sheep and a few other species (such as haemoglobin switching and involvement of a blood group substance in electrolyte transport through membranes) or result from the sheep being a suitable animal for surgical modification. Lymphatic and foetal studies in particular have been aided by the development of chronic surgical techniques.

Also under the heading of general biological research comes purification of sheep hormones and the use of sheep red cells in immunology. For reasons that are largely historical, the standard haemoglobin used for blood agar plates in bacteriology and the standard red cells for many immunological tests are from the sheep. The large volume of sheep endocrine organs available in abattoirs has permitted the isolation of many of the sheep hormones in pure form. Once their chemical structures were determined from the purified sheep hormones, some such as gonadotrophin releasing hormone and somatomedin have been chemically synthesized. Others, especially the anterior pituitary gland hormones, are too complex for laboratory synthesis but standard batches have been prepared from abattoir material by organizations such as the United States N.I.H. for supply to laboratories throughout the world.

Examples of these broad types of research are given in subsequent chapters of this book but it is illuminating to consider in the remainder of this chapter some areas of particular interest.

1.2 Foetal Research

The pregnant ewe and its lamb provide a close model for human pregnancy and have been studied extensively to provide insight into conditions affecting human development before birth. All of the other common experimental animals have very short gestation lengths, multiple births and/or very small foetuses. The ewe in contrast has normally only one or two lambs with births weights similar to those of human babies and a gestation length of about half that of the human baby. There are important differences in that the lamb at birth is more developed than the human baby and the limbs are a greater proportion of the body weight.

The initial serious research on the foetal lamb was done by Barcroft and his co-workers in Cambridge in the 1930s.[10] Barcroft was interested in characterizing the foetal environment and in particular in establishing how a foetus develops in an environment with a low oxygen tension. Although Barcroft performed some chronic surgery on lambs, most of his research and that of his immediate successors was done with acute preparations. It was in the 1960s that attention was given to techniques for surgical modification of the foetal lamb for chronic experimentation.[19] Use of these techniques has markedly extended knowledge of the foetal period and what can go wrong in it.

Several findings stand out as being of particular relevance to human medicine. Elucidation of the steps involved in the initiation of parturition in ewes (p. 65) showed that one step involved a rise in secretion of foetal adrenal cortical hormones in the last week of pregnancy. Experiments also done on pregnant ewes have shown that this rise in corticosteroids is important in lung maturation and production of pulmonary surfactant and this knowledge has had a dramatic impact on the treatment of respiratory distress in premature babies.

Persistent patency of the ductus arteriosus is also a problem in some new born babies and research on foetal lambs has shown that prostaglandins are probably the normal agents involved in its closure. Prostaglandin E_2 appeared to be the prostaglandin involved but since the discovery of prostacyclin, it is now believed that this or a similar unstable agent may be the active agent. Inhibitors of prostaglandin synthetase enzymes are used clinically for closure of the patent ductus in babies.

Another outcome of experiments on the initiation of parturition in the ewe was new insight into relaxation of the cervix prior to parturition. Liggins[24] tried to cervical route in ewes for administration of prostaglandin F_{2a}, a hormone which is produced during parturition. With the intracervical route, the cervix relaxed completely within 24 h and before uterine contractions had become marked. Subsequently he found that the cervix of non-pregnant ewes would relax similarly. Prior to this finding which may lead to a less exhausting delivery for women, it was believed that cervical relaxation occurred somehow due to the pressure waves produced by uterine contractions.

It has been known for many years that foetal sheep start "breathing" at about 30 days gestation and continue until about day 50 to breath when exteriorized in a saline bath.[4] After this time, breathing movements disappear in exteriorized lambs. Recently it was found by chronic recording techniques that foetal sheep breathe regularly *in utero* until birth. Breathing is depressed in the foetus by stress and it has been suggested that changes in foetal breathing may be an indicator of foetal stress.[9] Hopes of clinical use have yet to be realized.[23]

A prerequisite for successful intra-uterine research is the availability of satisfactory methods for sampling and recording. A new technique with considerable potential is one whereby a fistula is constructed between the uterus and the skin of a pregnant sheep so that the foetus can be repeatedly aseptically exteriorized into a special chamber for study.[27] During the period of exteriorization, the foetus remains connected via the umbilical cord to the placenta. After study, it is returned to the uterus and the fistula is closed.

1.3 Experimental Surgery

The sheep is an animal ideally suited to surgical modification for experimentation. Its size is convenient and somewhat similar to that of man. A comprehensive account of what can be done surgically to modify sheep for experimentation has been published separately[19] and hence a description of surgery on the sheep will not be given in this book. However the following is an indication of some of the techniques commonly used.

Translocation of organs is often a useful technique. The mobility of the left kidney allows this organ to be relocated with ease to a subcutaneous position where repeated biopsy is then made simple. More complicated operations produce similar subcutaneous translocations of the spleen and lengths of intestine. The thyroid gland can also be translocated keeping branches of the carotid artery and jugular vein intact for blood supply. If these blood vessels are then enclosed either together or separately in skin loops, a preparation is made in which the artery can be cannulated for perfusion of substances through the gland and the whole effluent from the gland (plus a small area of skin) can be collected.

Many of the blood vessels of the sheep are sufficiently large to allow vascular anastomoses to be performed relatively easily and this has led to the development of techniques whereby the adrenal gland, pancreas, ovary, ovary plus uterus, or uterus are autotransplanted to the neck with the main artery or arteries and vein being anastomosed to the carotid artery and jugular vein respectively. These vessels are then enclosed in skin loops producing preparations equivalent to that for the thyroid gland described above and which can be studied virtually as ex-vivo organs. Most of these preparations function indefinitely although for unknown reasons most pancreatic transplants cease to respond to appropriate stimuli after a few months.[3] Looking to the future, the development of techniques for easy production of identical twin lambs (p. 66) means that it will be possible to transplant other organs to obtain better access to the blood supply without the problem of rejection.

The greatest surgical attention has been given to the gastro-intestinal tract and this has been reviewed elsewhere.[19] The most common operation performed on sheep is the creation of a "cannulated fistula" into the rumen and several methods for such rumen cannulation are available. Other methods are available for cannulation of the intestine and parotid, bile and pancreatic ducts with either single or re-entrant cannulae. Animals with such surgical modifications live for many years in the laboratory and the presence of cannulae usually does not affect digestion, rates of passage or other parameters.

Another noteworthy area of surgery for which the sheep is ideally suited is lymphatic duct cannulation. It has been possible to cannulate afferent and efferent ducts of most of the major lymph nodes and to keep these cannulae patent for periods of up to three months. Ducts have also been cannulated in the foetus and have continued to produce lymph after birth of the lamb.

1.4 Biomedical Devices Research

A rapidly expanding field in several countries led by the United States, is the development of devices for temporary or permanent replacement of function of organs of the human body. Many devices have been developed or investigated using large laboratory animals including the sheep. Apart from problems associated with microsurgical techniques that would be required if the traditional small laboratory animals were used, the use of larger animals often means that the potential device can be developed in a size configuration that is suitable for eventual human use. These advantages outweigh the greater costs associated with the use of larger animals.

Possibly the most spectacular development in the field of artificial organs is the artificial heart. The sheep was used initially for research on this device as the thorax of a large sheep will accommodate an artificial heart suitable for implantation in a human.[2] Technical problems associated with early stages of development of artificial hearts resulted in sheep living for only a few hours. A switch was made to calves of about 45 kg body weight and gradually survival increased from hours to days, weeks and then months. A human size heart is suitable for a 45 kg calf but growth to 180 kg after 6 months results in the artificial heart in a calf becoming very inadequate. A change to another species is indicated and the sheep is the most obvious choice. The earlier developmental problems have almost certainly been overcome as half an artificial heart, the trans-apical left ventricular assist device, which has had recent clinical use in man, was largely developed in sheep and has worked in sheep with few problems for periods of up to a month.[32]

A second artificial organ is the artificial kidney which is used extracorporeally and was developed using dogs. The sheep has however been used successfully in artificial kidney research[34] and in the related research into haemodialysis.[36] The pattern of branching of the renal artery of the sheep inside the kidney allows the functioning renal mass to be reduced by 90% so as to reproduce the condition of partial renal failure.[13]

Extracorporeal membrane oxygenation (ECMO) is a development into longer-term use of bypass oxygenators used for intracardiac surgery. ECMO ideally will partially or totally replace lung function for weeks so as to allow time for healing of damaged and poorly functioning lungs. Most ECMO rsearch has been done with sheep[5,15,21] and oxygenators have functioned in sheep for 2 weeks and have been used successfully with humans.

A modification of ECMO is the "carbon dioxide lung" in which most of the carbon dioxide is removed from a small portion of the arterial blood.[16] This lowers the P_{CO_2} of arterial blood and so eliminates the carbon dioxide drive for breathing. If oxygen is delivered into the trachea, respiratory movements can be stopped indefinitely as oxygen will move to the alveoli by "paradoxical" respiration. This type of device would allow time for healing of damage to the thoracic structure or the lungs. There is little risk of pulmonary oxygen toxicity[22] if the nitrogen is not washed out of the lungs. One group has commenced development of longer-term membrane oxygenation and has actually constructed in sheep after hemipneumonectomy a skin-lined intrathoracic cavity of 2 to 3 litres which might accommodate an artificial device such as a permanent membrane oxygenator.[45]

Vascular prostheses have been tested in sheep[44] as have vascular access devices,[13] a passive artificial muscle,[20] a prosthetic urinary bladder replacement[39] and an artificial anus.[38] The sheep was considered a more satisfactory animal for testing of cardiac pacemakers than the dog.[11]

1.5 Other Research

A few other experiments on sheep presented below give other examples of the value of the sheep as a "large" experimental animal. When (during anaesthesia) burns were applied to 40% of the skin area, sheep were studied for several days.[43] These sheep ate and drank and showed no signs of discomfort due to the fact that the cutaneous nerve endings had been destroyed. A comment was made that the cardiovascular and renal parameters with the exception of the blood pH were comparable with man. The docility of the sheep has also allowed it to be used in long-term experiments when contusion injuries were done to the spinal cord in order to mimic paraplegic injuries.[47] Other long-term experiments have involved the investigation of changes in experimental arteriovenous fistulae[40] and aneurisms,[41] joint reconstruction,[46] cardiac valve replacements,[6] foetal growth retardation after chronic alcohol ingestion,[33] defects in enamel from trauma during tooth development,[42] experimental dermal cysts,[30] cartilage transplantation,[26] tissue biocompatability of experimental plastics,[17] and chronic effects of smoking.[28]

Of a shorter-term nature, experiments have been done with sheep on spinal anaesthesia,[1] decompression sickness,[37] computer-controlled muscle relaxation[8] and pulmonary air embolism.[12]

Numerous other examples of sheep research are given in subsequent chapters.

References

1. Adams, H. J. (1977). Bupivacaine: Morphological effects on spinal cords of cats and durations of spinal anesthesia in sheep. *Pharmacol. Res. Comm.*, **9**: 847–855.
2. Anderson, W. D. (1971). *Thorax of Sheep and Man; an Anatomy Atlas.* Dillon Press; Minneapolis.
3. Arcus, A. C., Beaven, D. W., Hart, D. S. and Holland, G. W. (1979). Long-term hormonal secretion from the autotransplanted sheep pancreas. *Diabetologia* **16**: 325–330.

4. Barcroft, J. and Barron, D. H. (1937). Movements in midfoetal life in the sheep embryo. *J. Physiol.*, **91**: 329–351.
5. Birek, A., Duffin, J., Glynn, M. F. X. and Cooper, J. D. (1976). The effect of sulfinpyrazone on platelet and pulmonary responses to onset of membrane oxygenator perfusion. *Trans. Am. Soc. artif. internal Organs*, **22**: 94–100.
6. Borrie, J. and Redshaw, N. R. (1973). Stented pulmonary valve allografts as tricuspid valve substitutes in sheep. *Thorax*, **28**: 102–106.
7. Bustad, L. K. and McClellan, R. O. (1966). *Swine in Biomedical Research*, Battelle Memorial Institute.
8. Cass, N., Brown, W. A., Ng, K. C. and Lampard, D. G (1980). Dosage patterns of nondepolarizing neuromuscular blockers in sheep. *Anaesth. Intens. Care*, **8**: 13–15.
9. Chapman, R. L. K., Dawes, G. S., Rurak, D. W. and Wilds, P. L. (1978). Intermittent breathing before death in fetal lambs. *Am. J. Obstet. Gynecol.*, **131**: 894–898.
10. Comline, R. S., Cross, K. W., Dawes, G. S. and Nathanielaz, P. W. (1973), editors. *Foetal and Neonatal Physiology: Proceedings of the Sir Joseph Barcroft Centenary Symposium*. Cambridge University Press; Cambridge.
11. Cummings, J. R., Gelok, R., Grace, J. L. and Salkind, A. J. (1973). Long-term evaluation in large dogs and sheep of a series of new fixed-rate and ventricular synchronous pacemakers. *J. Thoracic Cardiovas. Surg.*, **66**: 645–652.
12. Deal, C. W., Fielden, B. P. and Monk, I. (1971). Hemodynamic effects of pulmonary air embolism. *J. Surg. Res.*, **11**: 533–538.
13. Dennis, M. B., Cole, J. J., Jensen, W. M., Congdon, E. and Scribner, B. H. (1978). Effects of site and length on function of intravenous fistula catheters. *J. Surg. Res.*, **25**: 143–146.
14. Forbes, C. D. and Prentice, C. R. M. (1978). Thrombus formation and artificial surfaces. *Br. med. Bul.*, **34**: 201–207.
15. Fountain, S. W., Duffin, J., Ward, C. A., Osada, H., Martin, B. A. and Cooper, J. D. (1979). Biocompatability of standard and silica-free silicone rubber membrane oxygenators. *Am. J. Physiol.*, **236**: H371–H375.
16. Gattinoni, L., Kolobow, T., Tomlinson, T., White, D. and Pierce, J. (1978). Control of intermittent positive pressure breathing (IPPB) by extracorporeal removal of carbon dioxide. *Br. J. Anaesth.*, **50**: 753–758.
17. Gilding, D. K., Green, G. F., Annis, D. and Wilson, J. G. (1978). Soft tissue ingrowth hydrogels. *Trans. Am. Soc. artif. internal Organs*, **24**: 411–413.
18. Hackett, E. (1973). *Blood the Paramount Humour*. Jonathan Cape; London.
19. Hecker, J. F. (1974). *Experimental Surgery on Small Ruminants*. Butterworths; London.
20. Helmer, J. D. and Hughes, K. E. (1973). Implantable passive artificial muscle. *Trans. Am. Soc., artif. internal Organs*, **19**: 38[illegible]-384.
21. Kolobow, T., Tomlinson, T., Pierce, J. and [illegible]attinoni, L. (1976). Platelet response to long-term spiral coil membrane [illegible]ng bypass without

heparin using a carbon silicone rubber membrane. *Trans. Am. Soc. artif. internal Organs*, **22**: 110–117.

22. deLemos, R., Wolfsdorf, J., Nachman, R., *et al.* (1969). Lung injury from oxygen in lambs: The role of artificial ventilation. *Anesthesiology*, **30**: 609–618.
23. Lewis, P. and Boyland, P. (1979). Fetal breathing: A review. *Am. J. obstet. Gynecol.*, **134**: 587–589.
24. Liggins, G. C. (1978). Ripening of the cervix. *Semin. Perinatol.*, **2/3**: 261–271.
25. McDougall, D. S. A. (1976), editor. *British Sheep*, 4th edition. The National Sheep Association.
26. McKibbin, B. (1971). Immature joint cartilage and the homograft reaction. *J. Bone Joint Surg.*, **53B**: 123–135.
27. Maloney, J. E., Baird, J., Brodecky, V. and Dixon, J. (1978). A technique for the direct observation of the unanasthetized fetal sheep. *Am. J. Obstet. Gynecol.*, **131**: 281–285.
28. Mawdesley-Thomas, L. E. and Healey, P. (1973). Experimental bronchitis in lambs exposed to cigarette smoke. *Arch. Environ. Health*, **27**: 248–250.
29. *Michelin Guide: Paris* (1972), p. 134. John Murray (Publishers) Ltd; London.
30. Molyneux, G. S. and Lyne, A. G. (1961). Studies on experimental dermal cysts in sheep. *Aust. J. biol. Sci.*, **14**: 131–140.
31. Mount, L. E. and Ingram, D. L. (1971). *The Pig as a Laboratory Animal*. Academic Press; London and New York.
32. Peters, J. L., Fulkumasu, H., McRae, J. C., *et al.* (1978). Prolonged transapical left ventricular bypass (TALVB) in sheep and man. *Trans. Am. Soc. artif. internal Organs*, **24**: 113–120.
33. Potter, B. J., Belling, G. B., Mano, M. T. and Hetzel, B. S. (1980). Experimental production of growth retardation in the sheep fetus after exposure to alcohol. *Med. J. Aust.*, **2**: 191–193.
34. Richardson, P. D., Galletti, P. M. and Born, G. V. R. (1976). Regional administration of drugs to control thrombosis in artificial organs. *Trans. Am. Soc. artif. Internal Organs*, **22**: 22–28.
35. Ryder, M. L. and Stephenson, S. K. (1968). *Wool Growth*. Academic Press; London and New York.
36. Schmer, G., Teng, L. N. L., Cole, J. J., Vizzo, J. E., Francisco, M. M. and Scribner, B. H. (1976). Successful use of a totally heparin grafted hemodialysis system in sheep. *Trans. Am. Soc. artif. internal Organs*, **22**: 654–662.
37. Smith, K. H. and Spencer, M. P. (1970). Doppler indices of decompression sickness: Their evaluation and use. *Aerospace Med.*, **41**: 1396–1400.
38. Stanley, T. H., Kessler, J. R., Wiseman, L. R. and Blumle, C. A. (1973). Artificial control of the anal colostomy in sheep. *J. Surg. Res.*, **9**: 223–227.
39. Stanley, T. H., Feminella, J. G., Priestly, J. B., Lattimer, J. K. and Kessler, T. R. (1971). Bladder regeneration after cystectomy and prosthetic

urinary bladder replacement. *Trans. Am. Soc. artif. internal Organs*, **17**: 134–137.
40. Stehbens, W. E. (1968). Blood vessel changes in chronic experimental arteriovenous fistulas. *Surg. Gynec. Obstet.*, 127: 327–338.
41. Stehbens, W. E. (1979). Chronic changes in the walls of experimentally produced aneurisms in sheep. *Sheep, Gynec. Obstet.*, **149**: 43–48. (See also Stehbens, W. E. (1974). *Am. J. Path.*, **76**: 377–400.)
42. Suckling, G. (1980). Defects of enamel in sheep resulting from trauma during tooth development. *J. dental Res.*, **59**: 1541–1548.
43. Traber, D. L., Bohs, C. T., Carvajal, H. F., Linares, H. A., Miller, T. H. and Larson, D. L. (1979). Early cardiopulmonary and renal function in thermally injured sheep. *Surg. Gynec. Obstet.*, **148**: 753–758.
44. Trudell, L. A., Boudreau, L., Van De Water, J. M., Jauregui, H., Richardson, P. D. and Galletti, P. M. (1978). Alcohol treated PTFE vascular grafts. *Trans. Am. Soc. artif. internal Organs*, **24**: 320–323.
45. Trudell, L. A., Peirce, E. C., Teplitz, H. C., Richardson, P. D. and Galletti, P. M. (1979). A surgical approach to the implantation of an artificial lung. *Trans. Am. Soc. artif. internal Organs*, **25**: 462–465.
46. Wigren, A. and Olerud, S. (1973). Reincorporation of an avascular articular surface bearing bone fragment. *Upsala J. med. Sci.*, Suppl. 14.
47. Yeo, J., Payne, W. and Hinwood, B. (1975). The experimental contusion injury of the spinal cord in sheep. *Paraplegia*, **12**: 275–298.

Chapter 2

Anatomy

Each type of animal has anatomical peculiarities. In some, these are unique while in others they appear to be associated with particular life styles and are usually also found in animals with similar life styles. The sheep has anatomical features of both of these types. In this chapter, rather than attempt to provide a brief account of the anatomy of the sheep, attention is drawn to those anatomical features that tend to be unusual. It is anticipated that many readers will have medical backgrounds and it is assumed that they will make the necessary adjustments in terminology resultant from the sheep (like nearly all other laboratory animals) adopting a horizontal posture rather than the normal vertical posture of man.

Unfortunately no good comprehensive account of the anatomy of the sheep has been written. This is a pity as the sheep is used in dissection for teaching anatomy in several veterinary schools. The standard veterinary anatomical text is *The Anatomy of the Domestic Animals.*[9] Previous editions had little on the sheep but the latest edition has nearly 500 pages on ruminants of which many describe the sheep. There is also a dissection guide to the sheep[22] which contains comprehensive details on anatomy. Sizes quoted here are mainly from this guide. *Guide to the Dissection of Domestic Animals*[13] also provides many details. Several good anatomical diagrams of the sheep are provided by Nickel *et al.*[24] and Popesko.[25]

2.1 Skeleton and Muscles

The vertebral formula of the sheep is:

C 7, T 13 (12–14), L 6–7, S 4, Cy 16–18.

The digits are reduced to two main digits (cloven hoofed) with two rudimentary digits at the back of the metacarpus and metatarsus. Sheep are "cursorial" animals in that they stand on relatively straight legs.[17] Details of their gait at different speeds are given by Jayes and Alexander.[17] Sinus development in the skull is marked[15] and bones in the dorsal portion of the skull are heavily developed in breeds with horns. This is especially marked in rams. This development is much greater than in the goat and, with the associated differences in methods of fighting amongst males, has been used by archeologists to identify sheep skeletons in the presence of goat skeletons.[28]

The sternum consists of six fused sternebrae and is divided into a manubrium sterni, a body and a xiphoid process. The first eight ribs articulate with the sternum by their cartilages while the other ribs articulate by costal cartilages with the preceding cartilage. The first six ribs are directed cranially, ribs 7 and 8 are nearly vertical while the last ribs are inclined caudally.[2]

Accounts by Kresan[19] and Benevent[3] are given in German and French respectively for the muscles in the sheep.

2.2 Respiratory and Circulatory Systems

The thorax and its contents are illustrated and described by Anderson.[2] His account was written to identify differences between the thorax and its contents of man and sheep as an aid to the development of artificial hearts to be used in the thoracic cavity. The thorax is coned shaped, narrowest anteriorly and flattened laterally. Given that the thorax is compressed in a different plane to that in man and that the volume of abdominal contents has caused the diaphragm to protrude a considerable distance into the thorax, the similarities are considerable. Specific points of difference are that the lobes of the lung are different[14] (see fig. 1) and that the sheep has a separate right apical bronchus from the trachea supplying the right apical lobe. The hemiazygous vein is large and commonly drains into the coronary sinus while the small azygous vein drains into the anterior vena cava. Also an "anterior aorta" (brachiocephalic trunk) comes off from the arch of the aorta and branches in the apex of the thorax into a right and a left brachial artery and two common carotid arteries.

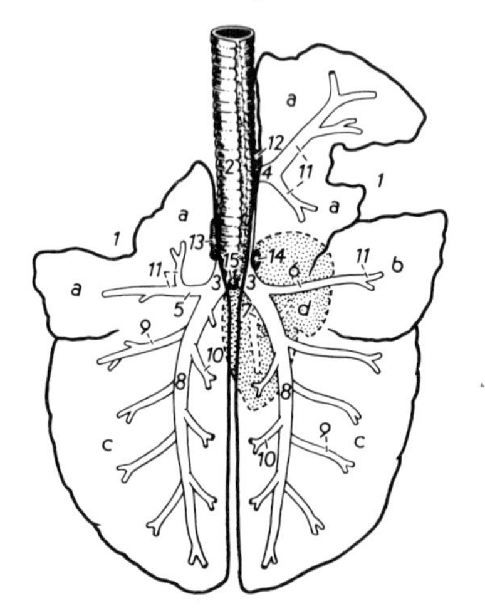

Fig. 1 *Lobulation bronchial tree and lymph nodes of the lung.*
a, Cranial lobe; b, Middle lobe; c, Caudate lobe; d, Accessory lobe (stippled). 1, Cardiac notch; 2, Trachea; 3, Principle bronchus; 4, Tracheal bronchus; 5, Cranial bronchus; 6, Middle bronchus; 7, Accessory bronchus; 8, Caudal bronchus; 9, First (left) and third (right) ventral segmental bronchi of caudal lobe; 11, Further segmental bronchi; 12, Cranial; 13, Left and 14, Right tracheobronchial lymph nodes.
[From Nickel, Schummer, Sieferle, Volume 11, *The Viscera of Domestic Animals*, 2. Royalty, Verlag Paul Parey, Berlin and Hamburg (1979).]

There is a common carotid artery on each side of the neck rather than separate internal and external carotid arteries. Chemo- and baro-receptors are present at the junction of the common carotid and occipital arteries. Carotid bodies as such are found there but there is no recognizable carotid sinus.[32] Blood supply to the brain is from the carotid, occipital and the vertebral arteries. Different from man is that both carotid arteries can suddenly be occluded without a sheep losing consciousness provided that the occipital arteries are intact.[32] In the skull, three branches from the internal maxillary artery form a carotid rete which is surrounded by venous sinuses and which acts as a heat exchange mechanism (p. 89). Anatomical details of the arterial supply to the appendages are given by Ghoshal and Getty[10,11] and of branches of the aorta by Kowatochev.[18] Peculiarities of the abdominal arteries are that there are usually separate branches from the aorta to the forestomachs and to the intestines, that the posterior mesenteric artery is small and that the spleen has a single pedicle. Naturally the arterial supply to the forestomachs and abomasum[1] is more complex than that to the stomach of man.

There is one (sometimes two) *os cordis* in the adult sheep heart between the atria and ventricles. This bone appears to provide support for the valves.[8]

2.3 Digestive System

It is the digestive system of the sheep that provides the greatest contrast to that of man and other experimental animals. The great volume of abdominal viscera comes as a surprise to most surgeons accustomed to operating on humans or cats and dogs. Much of this is the voluminous forestomachs although the intestines are also of considerable bulk. To support the weight of these viscera, the abdominal wall is comprised predominantly of fibrous tissue and the amount of muscle in the wall is relatively small except near the ribs and the lumbar vertebrae. The high proportion of fibrous tissue causes the abdominal wall to be comparatively thin (see Figs 2, 3 and 4).

The dental formula for the permanent teeth is:

$$2(\mathrm{I}\ 0/4,\ \mathrm{C}\ 0/0,\ \mathrm{P}\ 3/3,\ \mathrm{M}\ 3/3) = 32$$

while that for the temporary teeth is:

$$2(\mathrm{dI}\ 0/4,\ \mathrm{dC}\ 0/0,\ \mathrm{dP}\ 3/3,\ \mathrm{dM}\ 0/0) = 20.$$

As in the cow and other ruminants, incisors on the upper jaw are absent and instead there is a dental pad against which the lower incisors act to shear off grass. Premolar and molar teeth are flattened for grinding with sharp edges laterally on teeth in the upper jaw and medially on those in the bottom jaw. The tongue is large and rough.

The usual method for determining the age of sheep is from the teeth; usually from the incisors. There are some differences which are due to breed and nutrition[7] but in general the following[22] applies to permanent teeth:

M_1 — 3 months
M_2 — 9 months
I_1 — 10–12 months (one toothed)
M_3 — 18 months
I_2, Pm_1, Pm_2 — 18–24 months (two toothed)
I_3, Pm_3 — 2–2·5 years (three toothed)
I_4 — 3–4 years (four toothed)

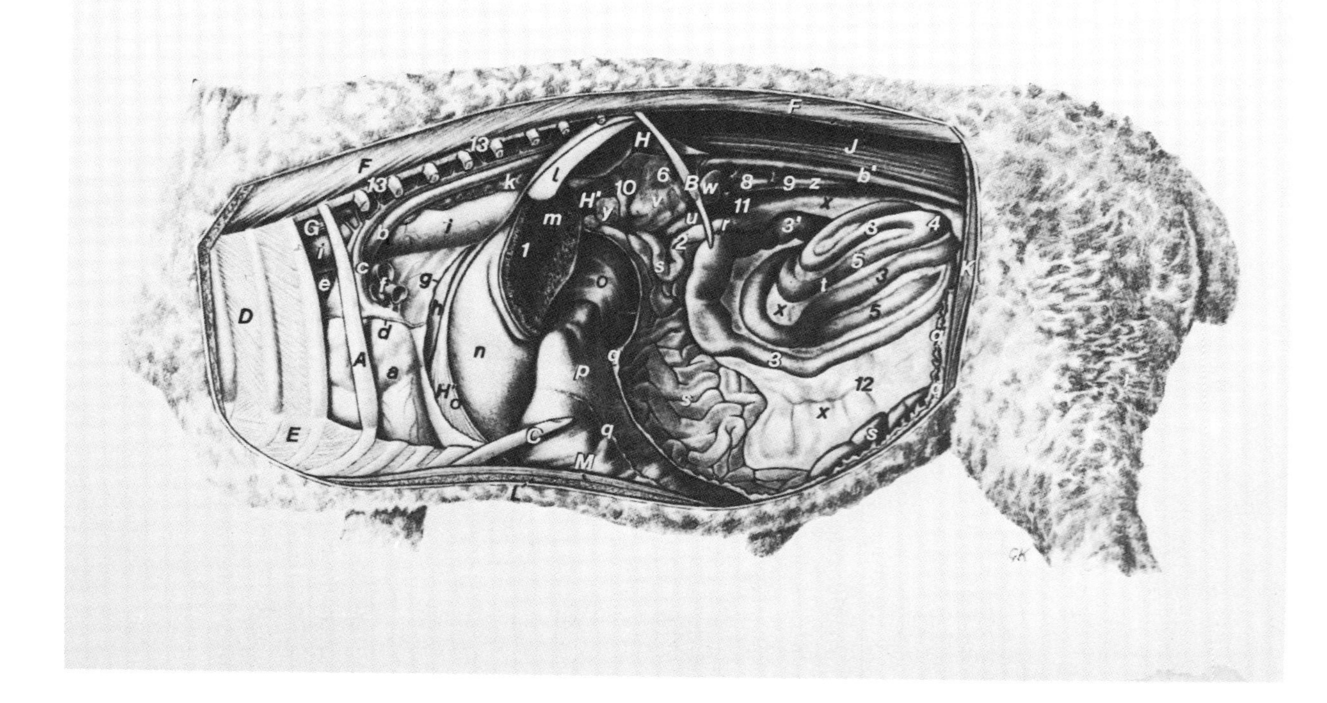

F
13
H
J
b'
k
l
10
6
B
w
8
9
z
H'
y
v
u
11
r
x
G
m
b
i
2
3'
4
1
s
c
o
t
K
e
f
g
h
D
d
n
q
p
A
a
12
H''
C
E
M
L

Fig. 2 *Topography of the thoracic and abdominal organs. Left aspect.*

The left lung, left half of diaphragm, costal arch, left kidney, and most of the rumen have been removed.

A, Fourth rib; B, Thirteenth rib; C, Eighth costal cartilage; D, Ext. internal muscle; E, Int. intercostal muscle; F, Longissimus; G, Longus colli; H, Diaphragm, its left crus; H′, part of its right crus; H″, its sternal part; J, Psoas musculature; K, Int. and ext. abdominal oblique muscles; L, Pectoral muscles; M, Rectus abdominis.

a, Heart inside pericardium; b, b′, Aorta; c, Left azygous vein; d, Left phrenic nerve; e, Trachea; f, Root of left lung; g, Caudal mediastinum, its cut edge; h, Accessory lobe of right lung; i, Oesophagus; k, Caudal mediastinal lymph node; l, Spleen; m, Rumen, part of its cranial sac; n, Reticulum; o, Omasum; p, Abomasum; q, Supf. wall; q′, deep wall of greater omentum; r, Ascending duodenum; s, Jejunum; t, Spiral loop of ascending colon; u, Descending colon; v, Pancreas; w, Left adrenal gland; x, Mesentery; y, Ruminal lymph nodes; z, Caudal vena cava.

1, Rumenoreticular fold; 2, Duodenojejunal flexure; 3–5, On the ascending colon: 3′, Distal part of proximal loop; 3, centripetal turns; 4, central flexure; 5, centrifugal turns; 6, Celiac artery; 7, Cranial mesenteric artery; 8, Right renal artery; 9, Left renal artery; 10, Splenic vein; 11, Left renal vein; 12, Jejunal vessels; 13, Dorsal intercostal artery and vein.

[From Nickel, Schummer, Sieferle, Volume 11, *The Viscera of Domestic Animals*, 2. Royalty, Verlag Paul Parey, Berlin and Hamburg (1979).]

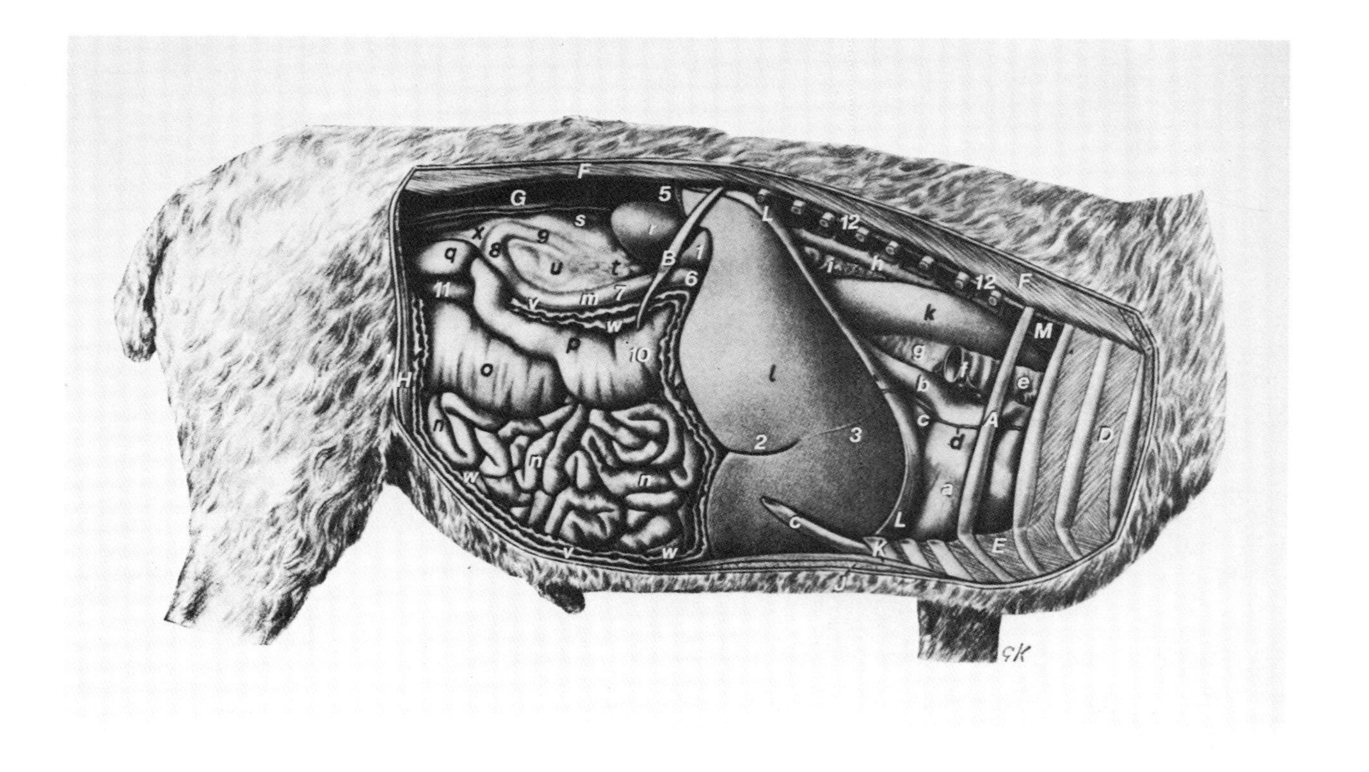
F
G
s
5
r
L
12
x
g
q
8
u
t
B
1
i
h
11
6
7
m
v
12
F
w
k
M
p
10
g
f
H
o
l
b
e
c
A
n
2
3
d
D
n
n
w
a
c
L
K
E
v
w
J
GK

Fig. 3 *Topography of the thoracic and abdominal organs. Right aspect.*

The right lung, right half of diaphragm, costal arch, and parts of the greater omentum have been removed.

A, Fourth rib; B, Thirteenth rib; C, Eighth costal cartilage; D, Ext. intercostal muscle; E, Int. intercostal muscle; F, Longissimus; G, Psoas musculature; H, Int. and ext. abdominal oblique muscles; J, Pectoral muscles; K, Rectus abdominus; L, Diaphragm; M, Longus colli.

a, Heart inside pericardium; b, Caudal vena cava; c, Plica venae cavae; d, Right phrenic nerve; e, Trachea with tracheal bronchus; f, Root of right lung; g, Caudal mediastinum; h, Aorta; i, Caudal mediastinal lymph node; k, Oesophagus; l, Liver; m, Duodenum; n, Jejunum; o, Caecum; p, Ascending colon; q, Descending colon; r, Right kidney; s, Right ureter; t, Pancreas; u, Mesoduodenum; v, Superficial wall; w, Deep wall of greater omentum; x, Duodenocolic fold

1, Caudate process of liver; 2, Notch for round ligament; 3, Falciform ligament; 4, Gall bladder; 5, Right triangular ligament; 6, Cranial duodenal flexure; 7, Descending duodenum; 8, Caudal duodenal flexure; 9, Ascending duodenum; 10, Ventrolateral part of the proximal loop of the ascending colon; 11, First centripetal turn of spiral colon; 12, Dorsal intercostal artery and vein.

[From Nickel, Schummer, Sieferle, Volume 11, *The Viscera of Domestic Animals*, 2. Royalty, Verlag Paul Parey, Berlin and Hamburg (1979).]

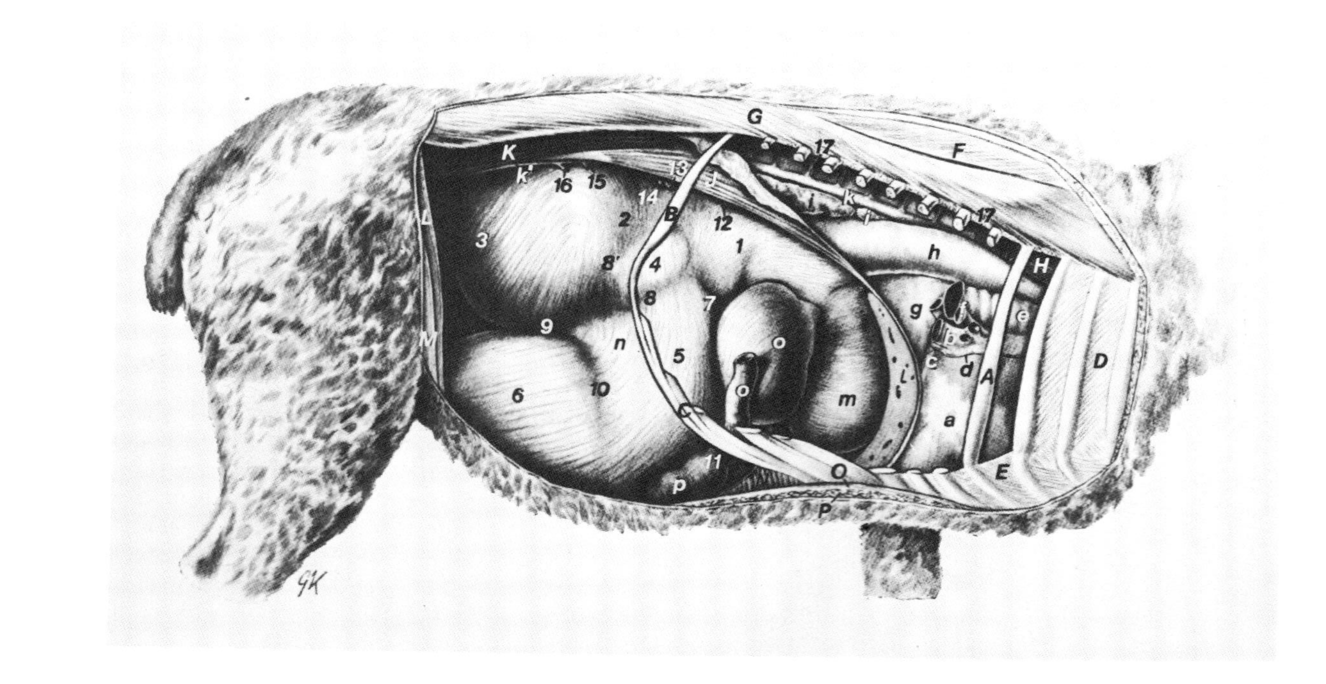

G
17
F
K
k'
16
15
13
J
14
i
k
B
12
2
3
1
h
8'
4
H
8
7
g
f
e
9
o
M
n
5
b
c
d
D
A
6
10
o
m
C
a
11
O
E
p
P
GK

Fig. 4 *Topography of the thoracic and abdominal organs. Right aspect.*

The right lung, right half of diaphragm, omenta, intestines, kidneys, and most of the liver have been removed.

A, Fourth rib; B, Thirteenth rib; C, Costal arch; D, Ext. intercostal muscles; E, Int. intercostal muscles; F, Spinalis et semispinalis thoracis et cervicis; G, Longissimus; H, Longus colli; J, Right crus of diaphragm; K, Psoas musculature; L, Int. abdominal muscle; M, Ext. abdominal oblique muscle; N, Transversus abdominis; O, Rectus abdominis; P, Pectoral muscles.

a, Heart inside pericardium; b, Caudal vena cava; c, Plica venae cavae; d, Right phrenic nerve; e, Trachea; f, Root of right lung; g, Caudal mediastinum; h, Oesophagus; i, Caudal mediastinal lymph nodes; k, k′, Aorta; l, Liver; m, Reticulum; n, Rumen; o, Omasum; p, Abomasum; q, Duodenum.

1, Cranial sac of rumen; 2, Dorsal sac of rumen; 3, Caudodorsal blind sac; 4, Insula ruminis; 5, Ventral sac of rumen; 6, Caudoventral blind sac; 7, Cranial groove; 8, 8′, Right longitudinal groove; 9, Caudal groove; 10, Ventral coronary groove; 11, Pylorus; 12, Splenic artery; 13, Celiac artery; 14, Cranial mesenteric artery; 15, Right renal artery; 16, Left renal artery; 17, Dorsal intercostal artery and vein.

[From Nickel, Schummer, Sieferle, Volume 11, *The Viscera of Domestic Animals*, 2. Royalty, Verlag Paul Parey, Berlin and Hamburg (1979).]

The only method for the aging of older sheep is from growth lines in the cementum of the central incisor.[26] The detailed anatomy of the teeth of the sheep is described by Weinrib and Sharov.[34]

Swallowed food passes into the reticulo-rumen. The reticulum and the rumen are connected by a large opening and digesta passes readily between them. Often the reticulo-rumen is referred to simply as the rumen. These two compartments have different surfaces; the rumen is fully covered with papillae while the reticulum has papillae only in a hexagonal pattern which has given rise to the lay term of "honey-comb". Pillars partially divide the rumen into blind sacs of which the largest are the posterior dorsal and posterior ventral sacs. Between the reticulum and the next compartment, the omasum, is a small orifice, the reticulo-omasal orifice, which acts as a sphincter. Prominent on the dorsal wall of the reticulum are two lips, 10–12 cm long, which run from the oesophagus to the reticulo-omasal orifice. This is the "oesophageal groove" or "oesphageal-omasal sulcus". In young sheep, this groove is reflexly closed during suckling to divert milk directly into the abomasum as the oesophageal groove is continued through the last compartment of the forestomachs, the omasum. In young ruminants taught to drink directly from a bucket, the reflex rarely operates and so milk passes into the rumen where it is subjected to microbial attack.

The omasum is small and tightly packed with about seven large muscular folds and many smaller second and third order folds and has the lay term of "bible". The abomasum or true stomach also has about seven large mucosal folds but is otherwise similar to the stomach of the monogastric animal. Apart from the pyloric region, all of the mucosa secretes gastric juice.

The small intestine is long (Table 4) but is otherwise not unusual. It is suspended at the free edge of the mesentary which also suspends the large intestine (Fig. 5). The caecum is large and is continuous with the proximal colon. This narrows towards the spiral colon which consists of coils of centripetal colon and then of centrifugal colon.[29] After these coils follows a further length of colon which terminates at the rectum.

The liver has two main lobes, left and right, and a smaller third lobe. There is a gall bladder about 10 cm long (Fig. 6). The common bile duct is joined by the pancreatic duct and enters the duodenum at a distinct papilla of Valli.

Kidneys are devoid of external lobulation. The right kidney is

Table 4 Sizes of some abdominal viscera (from [22]).

Organ	Size
Rumen	8·81
Reticulum	0·71
Omasum	0·41
Abomasum	1·41
Small intestine	18–35 m
Caecum	0·25 m
Large intestine	4·5 m
Liver	0·7 kg

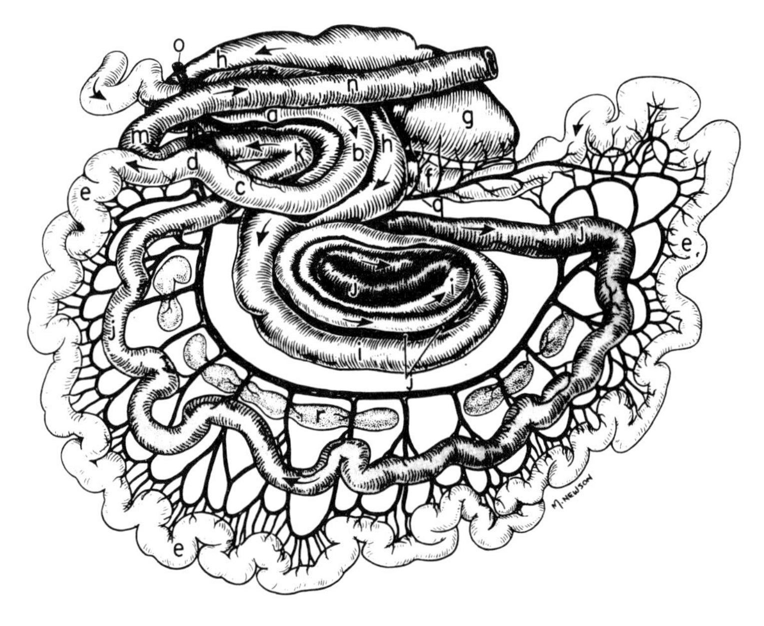

Fig. 5 *Anatomy of the intestine (as viewed from the left).*
a, Descending duodenum; b, Caudal flexure of duodenum; c, Ascending duodenum; d, Duodenal flexure; e, Jejunum; f, Ileum; g, Caecum; h, Proximal loop of colon; i, Centripetal gyri; j, Centrifugal gyri; k, Distal loop of colon; m, Transverse colon; n, Descending colon; o, Cranial mesenteric artery; q, Ileocaecal branch of ileocaecal artery; r, Mesenteric lymph nodes.
[Reproduced from Habel by permission of the author and publisher.]

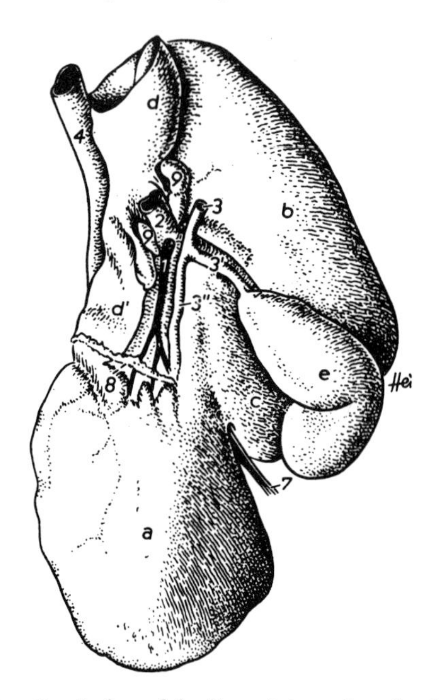

Fig. 6 *Lobes of the liver (visceral surface).*
a, Left Lobe; b, Right lobe; c, Quadrate lobe; d, Caudate lobe, its caudate process; d′, its papillary process, e, Gall bladder.
1, Hepatic branch of hepatic artery; 2, Portal vein; 3, Bile duct; 3′, Cystic duct; 3″, Common hepatic duct; 4, Caudal vena cava in its groove; 5, Left triangular ligament; 7, Falciform ligament; 8, Stump of lesser omentum; 9, Hepatic lymph nodes.
[From Nickel, Schummer, Sieferle, Volume 11, *The Viscera of Domestic Animals*, 2. Royalty, Verlag Paul Parey, Berlin and Hamburg (1979).]

always located in the right dorsal part of the abdomen against the liver. The presence of the large bulk of abdominal viscera has caused the left kidney to be variable in location. The ureters have small diameters.

2.4. Reproductive System

The cervix of the female reproductive tract is about 4 cm long and is narrow with convoluted folds. As a result of its narrow lumen and folds, catheterization of the uterus via the cervix is rarely possible. The body of the uterus is short (about 2 cm as measured in the lumen) but there are two long semi-coiled horns. Prominent on the internal surface of the uterus and its horns are rows of cotyledons where placental attachments occur in pregnancy. Ovaries are small (2–3 g) and the ovarian openings of the Fallopian tubes are inconspicuous.

The female reproductive tract is supplied with blood by the uterine arteries, the small cervical arteries and the ovarian arteries which are also small.[30] Venous drainage is via the uterine and ovarian veins which join to form the utero-ovarian vein.[30] The ovarian artery is interesting as it is highly coiled and is very closely adherent to the utero-ovarian vein and the ovarian vein.[20] This intimate adherence has physiological significance in the control of the life of the corpus luteum (p. 64).

In the male, testicles are large (200–300 g each) and are suspended in a large dependent scrotum. The testicular artery is compressed and surrounded by a venous plexus on top of the testis (p. 67). The cremaster muscle allows the testicles to be held close to the abdomen in cold or nearly to touch the ground in hot weather. The glans of the penis ends in a fine erectile urethral process which is about 4 cm long.

2.5 Skin and Wool

Surface areas of skin are given in Table 5.

Table 5 Areas of parts of the skin and formulae relating area to body weight (B) for Merino sheep (from Bennett[4]).

Region	Area
Lower forelegs	0·036 m^2
Lower hindlegs	0·048 m^2
Upper forelegs	0·060 m^2
Upper hindlegs	0·101 m^2
Ears	0·021 m^2
Head (less ears)	0·083 m^2
Neck	0·139 m^2
Trunk	0·579 m^2
Total area	1·067 m^2
Weight	38·8 kg
Area (m^2)	$= 0{\cdot}489 + 0{\cdot}015\,B$ (kg)
	$= 0{\cdot}171\,B^{0{\cdot}503}$ (kg)
	$= 0{\cdot}094\,B^{0{\cdot}67}$ (kg)

The woolled skin of the sheep is relatively thin with a free lipid layer of about 9 μm, a loose stratum of 16 μm and an uncornified epidermis of 17 μm.[21] On non-woolled areas it is thicker. Under the dermis, the subcutaneous fat layer is also thin and under this fat is the thin layer of cutaneous muscle which causes shivering in the cold.

The presence of wool follicles is the unusual feature of the sheep's skin. At about 60 days of gestational life, individual primary wool follicles form. About 10 days later, two second primary follicles appear, one on either side of the original primary follicle so as to form a triad of primary wool follicles. The primary follicles which will have grown a fibre by the time of birth each have an erector muscle (which is apparently non-functional) as well as a sudiferous (sweat) gland and a sebaceous (suint) gland.

Secondary wool follicles start forming to one side of each triad of primary follicles at about 90 days of gestation. These secondary follicles each have a sebaceous gland but have neither a sudiferous gland nor an erector muscle. The secondary follicles usually do not have a fibre projecting from the skin at birth. In most breeds at birth, the secondary to primary (S:P) follicle ratio is about 2–3:1. In many breeds, more secondary follicles bud off until 6 months of age so that the S:P ratio at maturity in most British breeds is 3–6:1 while in the Merino it is 20:1 or greater (up to 40:1). The significance of the S:P ratio is that the secondary fibres are finer than the primary fibres and hence a high ratio increases the overall fineness of the fleece. In

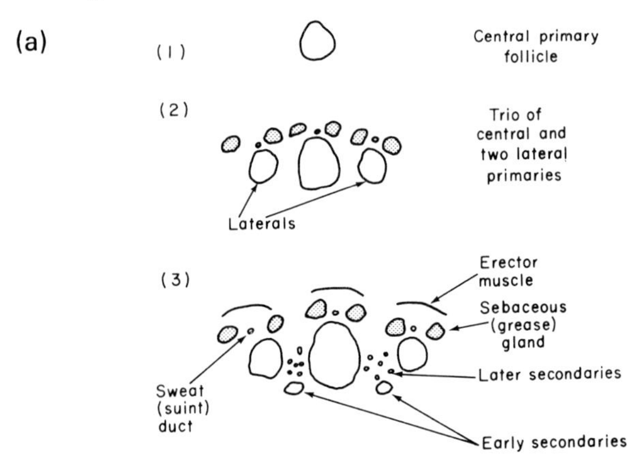

Fig. 7 *Diagrammatic representation of wool follicles.*
(a) Stages in follicular development.

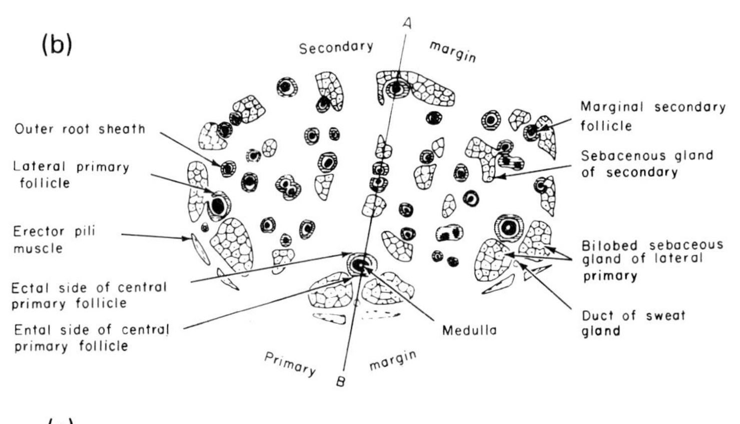

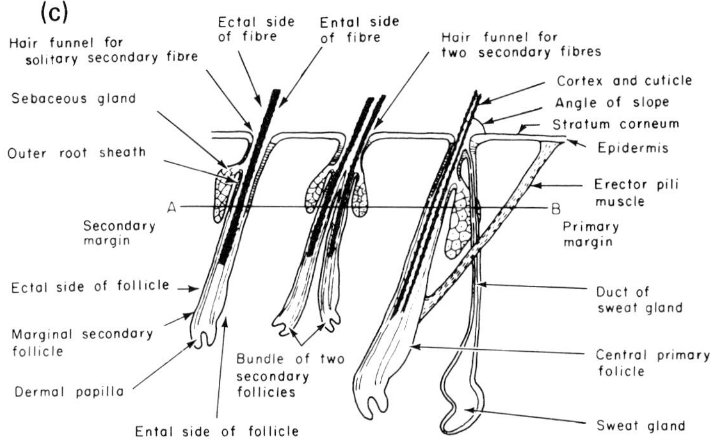

Fig. 7 *Diagrammatic representation of wool follicles*—continued
(b) Transverse section through an adult Merino wool follicle group at the level of the sebaceous gland.
(c) Longitudinal section through follicles in a plane passing through A–B in (b).
[From Ryder and Stephenson by permission of Academic Press Ltd.]

addition, the diameter of secondary fibres varies between breeds, being finest in the Merino (Fig. 7).

As well as wool fibres, sheep have kemps: coarse (100 μm or greater) short fibres which are shed naturally. They also have hair fibres which are intermediate between true wool and kemp fibres. Further details on skin and wool are provided by Ryder and Stephenson,[27] and Yeates *et al.*[36]

2.6. Other Anatomical Features

A dissection guide to the brain has been published by Briggs.[5]

As well as the normal chain of lymph nodes in the abdominal cavity, there are several small red nodes called haemal nodes which contain numerous red cells. Little is known of their function.[23] Of the peripheral lymph nodes, the popliteal, prefemoral and prescapular nodes are the most prominent. The pattern of the superficial lymph drainage is shown in Fig. 8 while details of the lacrimal drainage system are given by Gilanpour.[12]

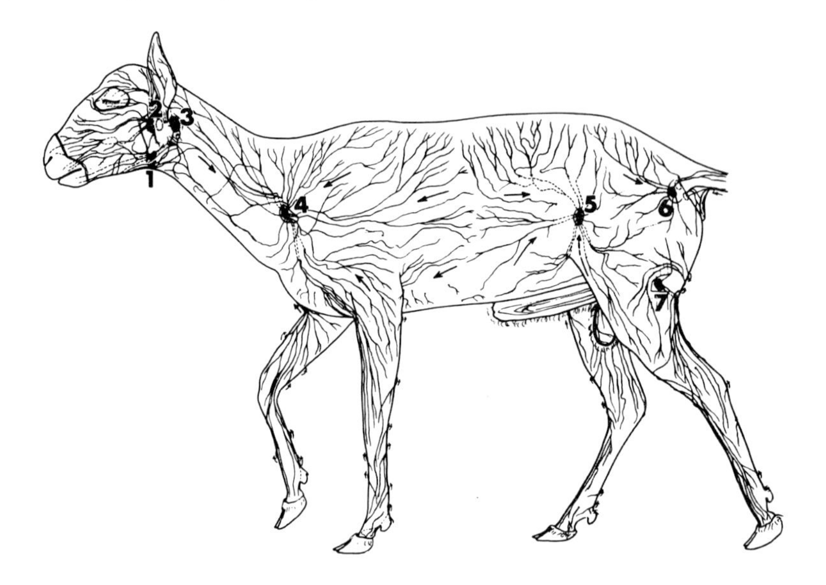

Fig. 8 *Superficial lymph flow and nodes.*
1, Mandibular; 2, Parotid; 3, Lateral retropharyngeal; 4, Superficial cervical; 5, Subiliac; 6, Gluteal, and 7, Popliteal nodes. Arrows indicate direction of lymph flow.
[From Getty by permission. After H. Grau, 1933.)

The pituitary gland is large. Accessory adrenal medullary tissue is occasionally found in the abdominal cavity.[31]

Agranular and glucagon-secreting cells are the most common cells in the islets of Langerhans of the pancreas. Insulin-secreting cells comprise only about 12% of islet cells. The islets are small.[35] There are the usual two pairs of parathyroid glands which are located near the submaxillary salivary gland and the thyroid gland.[6]

Adipose tissue probably undergoes cellular division as well as hypertrophy until about 11 months after which time fat deposition is only by hypertrophy. There are approximately 9×10^9 fat cells per carcase.[16]

References

1. Anderson, W. D. and Weber, A. F. (1969). Normal arterial supply to the ruminant (ovine) stomach. *J. anim. Sci.*, **28**: 379–382.
2. Anderson, W. D. (1971). *Thorax of Sheep and Man: an Anatomy Atlas.* Dillon Press; Minneapolis.
3. Benevent, M. (1968). Guide pour la dissection et l'identification des principaux muscles de la carcasse chez le mouton. *Ann. Biol. anim. Bioch. Biophys.*, **8**: 147–193.
4. Bennett, J. W. (1973). Regional body surface area of sheep. *J. agric. Sci.*, **81**: 429–432.
5. Briggs, E. A. (1946). *Anatomy of the Sheep's Brain: a Laboratory Atlas for Students of Zoology*, 2nd edition. Angus & Robertson: Sydney.
6. Calislar, T. and St. Clair, L. E. (1974). Parathyroid gland of domesticated ruminants. *J. Dairy Sci.*, **57**: 1263–1266.
7. Franklin, M. C. (1950). *The Influence of Diet on Dental Development in the Sheep.* Bulletin No. 252, Comm. sci. industr. Res.; Melbourne, Australia.
8. Frink, R. J. and Merrick, B. (1974). The sheep heart: Coronary and conduction system anatomy with special reference to the presence of an Os Cordis. *Anat. Rec.*, **179**: 189–200.
9. Getty, R. (1975). *Sisson and Grossman's the Anatomy of the Domestic Animals*, 5th edition. W. B. Saunders Co.; Philadelphia.
10. Ghoshal, N. G. and Getty, R. (1968). The arterial supply to the appendages of the sheep (*Ovis aries*). *Iowa St. J. Sci.*, **42**: 215–229.
11. Ghoshal N. G. and Getty, R. (1970). Comparative morphological study of the major arterial supply to the thoracic limb of the domestic animals (*Bos taurus, Ovis aries, Capra hircus, Sus scrofa domestica, Equus caballus*). *Anat. Anz.*, **127**: 422–443.
12. Gilanpour, H. (1979). Anatomic and radiographic studies of the lacrimal drainage system in sheep (*Ovis aries*). *Am. J. vet. Res.*, **40**: 1177–1179.
13. Habel, R. E. (1970). *Guide to the Dissection of Domestic Ruminants.* Published by the author; Ithaca, New York.
14. Hare, W. C. D. (1955). The broncho-pulmonary segments in the sheep. *J. Anat.*, **89**: 387–402.
15. Heyne, K. von, and Schumacher, G. H. (1967). Biometrische Untersuchungen an den Nebenhohlen der Nase von Ovis aries. *Anat. Anz.*, **120**: 433–443.

16. Hood, R. L. and Thornton, R. F. (1979). The cellularity of ovine adipose tissue. *Aust. J. agric. Res.*, **30**: 153–161.
17. Jayes, A. S. and Alexander, R. McN. (1978). Mechanics of locomotion of dogs (*Canis familiaris*) and sheep (*Ovis aries*). *J. Zool.*, **185**: 289–308.
18. Kowatschev, G. von (1968). Uber die Variabilitat der Aste der Brust- und Bauchaorta bei Schaffoten. *Anat. Anz.*, **122**: 37–47.
19. Kresan, J. von (1970). Beitrag zur Anatomie der Skelettmuskulatur des Schafes und der Ziege: Muskeln des Kopfes und des Stammes. *Anat. Anz.*, **126**: 38–58.
20. Lee, C. S. and O'Shea, J. D. (1976). The extrinsic blood vessels of the ovary of the sheeep. *J. Morphol.*, **148**: 287–303.
21. Lloyd, D. H., Amakiri, S. F. and McEwan Jenkinson, D. (1979). Structure of the sheep epidermis. *Res. vet. Sci.*, **26**: 180–182.
22. May, N. S. D. (1970). *The Anatomy of the Sheep: a Dissection Manual.* University of Queensland Press; Brisbane.
23. McIntosh, G. H. and Morris, B. (1971). The lymphatics of the kidney and the formation of renal lymph. *J. Physiol.*, **214**: 365–376.
24. Nickel, R., Schummer, A. and Seiferle, E. (1973). *The Viscera of the Domestic Animals*, translated by Sack, W. O. Verlag Paul Parey; Berlin.
25. Popesko, P. (1978). *Atlas of Topographical Anatomy of the Domestic Animals*. Saunders; Philadelphia.
26. Rudge, M. R. (1976). Ageing domestic sheep (*Ovis aries* L.) from growth lines in the cementum of the first incisor. *N.Z. J. Zool.*, **3**: 421–424.
27. Ryder, M. L. and Stephenson, S. K. (1968). See reference 35 in Chapter 1.
28. Schaffer, W. M. and Reed, C. A. (1972). The co-evolution of social behavior and cranial morphology in sheep and goats (*Bovidae, Caprini*). *Fieldiana Zool.*, **61**: 1–88.
29. Smith, R. N. (1955). The arrangement of the ansa spiralis of the sheep colon. *J. Anat.*, **89**: 246–249.
30. Tanudimadja, K., Getty, R. and Ghoshal, N. G. (1968). Arterial supply to the reproductive tract of the sheep. *Iowa St. J. Sci.*, **43**: 19–39.
31. Thwaites, C. J. (1971). Accessory adrenal glands in the sheep. *Aust. vet. J.*, **47**: 178.
32. Waites, G. M. H. (1960). The influence of the occipito-vertebral anastomoses on the carotid sinus reflex of the sheep. *Quart. J. exp. Physiol.*, **45**: 243–251.
33. Webb, R. M. (1944). The lymph nodes of the head and neck in the domestic animals. *Aust. vet. J.*, **20**: 181–188.
34. Weinreb, M. M. and Sharov, Y. (1964). Tooth development in sheep. *Am. J. vet. Res.*, **25**: 891–908.
35. White, A. W. and Harrop, C. J. F. (1975). The islets of Langerhans of the pancreas of Macropodid marsupials: A comparison with eutherian species. *Aust. J. Zool.*, **23**: 309–319.
36. Yeates, N. T. M., Edey, T. N. and Hill, M. K. (1975). *Animal Science: Reproduction, Climate, Meat, Wool.* Pergamon Press; Sydney.

Chapter 3

Physiology and Genetics

For each species of animal, there are aspects of its physiology which are similar to general mammalian physiology and also aspects that are different. It is these similarities and differences which mainly determine if a particular species can be used in experiments as a model for other species. The species which is usually modelled is *Homo sapiens* or man. The plan of this chapter is to give representative values for a range of physiological parameters for sheep in tables and to identify in the text those features which differ significantly or unexpectedly from the general mammalian pattern. Most of the parameters have been measured many times. The values presented have mostly been obtained in more recent experiments and references cited will usually lead the interested reader to previous work.

Physiology parameters for sheep in the literature have been expressed in a variety of terms including per unit body weight, per unit body weight raised to a power of between 0·66 and 0·75 (i.e. per unit of "metabolic weight"), per square metre of surface area, in relation to a cardiac index, or in absolute terms. There is little justification for expressing values in terms of surface area or cardiac index; these having been derived for comparisons between humans with formulae for humans usually being applied directly to sheep. Expression in terms of body weight$^{0·66–0·75}$ is more valid, especially for comparisons between species or between sheep of different sizes or different ages. For comparisons between sheep of the same age and breed, it is preferable to use absolute terms as variations in degrees of fatness, gut fill and wool covering may be greater than variations in lean body mass. In this chapter, most values will be given either directly or in relation to body weight.

3.1 Cardiovascular Physiology

For most cardiovascular parameters, the sheep is similar to man if allowance is made for the smaller weight of the average sheep. Possibly the most interesting difference is in the functioning of the spleen. Animal spleens tend to be either of a defensive type rich in lymphoid tissue or of a storage type. The human spleen is of the former type. Although the ruminant spleen from its histology is considered to be of an intermediate type,[358] little is known of the immunological function of the sheep's spleen. However its storage capacity is impressive as it can store up to one fourth of the red cell volume.[372] These stored cells (which tend to be younger than the average circulating red cell[379]) can be released into the circulation within about 3 min[114] and this can suddenly increase the circulating red cell mass by up to 38%[379]. The spleen of an exsanguinated sheep weighs about 45 g while that removed from an anaesthetized sheep weighs up to 600 g.[372] The difference lies in stored blood which may be in two compartments. One which drains freely from the spleen has an approximately normal haematocrit while the other which can be expelled from an excised spleen by an intrasplenic injection of adrenaline can have a haematocrit of up to 90%.[186] When a sheep is in an undisturbed state such as at night, the haematocrit is low.[114]

In man and many other animals, the Purkinje fibres of the heart cause ventricular contraction to be initiated at the apex from where it passes to the base. The walls of the sheep's heart are unusually richly supplied with Purkinje fibres except at the base and hence contraction occurs in an apico-basilar direction.[166] The Purkinje fibres are prominent on the ventricular wall and, because of the ease with which they can be obtained, they have been much studied.

The distribution of the cardiac output in the resting sheep is shown in Table 6. During exposure to heat, there is a redistribution of cardiac output and an increase in blood flow through arterio-venous anastomoses from 1·5 to 11% of the cardiac output.[160]

Denervation of the section of the carotid artery near its junction with the occipital artery which has baro- and chemo-receptors (p. 16) results in depression of respiration as shown by a rise in P_ACO_2, a fall in P_AO_2 and a rise in blood pressure and heart rate. These changes wear off after several weeks but the denervated animal still lacks a typical respiratory response to cyanide.[162] Vascular volume receptors are

Table 6 Blood flow to various organs (data from microspheres[160]).

Organ	Blood flow (ml/100 g tissue/min)	Per cent cardiac output
Leg skin	7	0·77
Ear skin	8	0·06
Body skin	14	10
Skeletal muscle	6	14
Thyroid gland	195	0·19
Adrenal gland	158	0·14
Kidneys	443	13·96
Spleen	198	4·66
Heart	161	7·72
Rumen	72	10
Small intestine	128	11
Large intestine	59	9
Brain	63	2·05
Spinal cord	13	0·17
Cardiac output	12·1	3·80 l/min

located in the left atrium as sheep increase their water intake to restore blood volume in proportion to water loss except when the left atrial appendage is crushed.[416]

Some published values for cardiovascular parameters are shown in Table 7. Both feeding and excitement increase heart rate and some of the heart rate values may be high because of these factors. The rise in heart rate with feeding is due largely to a direct effect of increased levels of angiotensin II having an inhibitory effect on vagal control of the heart.[233]

Fasting decreases the heart rate and values as low as 30 beats per min have been recorded after 4–5 days fasting.[391] For such low rates, the animal must be in a comparatively warm environment as a fasted animal has a lower metabolic heat production and hence its thermoneutral zone is increased (p. 87).

It is not generally realized that lymph ducts actively contract to propel lymph. In sheep, contractions occur 1–20 times per min and generate pressures up to 15 mm Hg.[66]

The sheep is possibly more sensitive to the hypertensive effect of partial constriction of a renal artery than other species.[288] Hypertension

Table 7 Cardiovascular parameters.

Parameter	Value
Heart rate (beats/min)	66,[162] 88,[283] 92,[357] 129[375]
Intrinsic heart rate (per min)	120[171]
Maximum heart rate (per min)	260–280[171]
ECG P wave duration (s)	0·054[340]
ECG P-R interval (s) (heart rate 84)	0·110[340]
ECG S-T segment (s) (heart rate 84)	0·192[340]
ECG Q-T interval (s) (heart rate 84)	0·315[340]
ECG Q-R-S duration (s) (heart rate 84)	0·044[340]
Pre-injection period (s) (heart rate 98)	0·076[39]
Left ventricular ejection time (s) (heart rate 98)	0·234[39]
Left ventricular contraction (d_p/d_t) (mm Hg/s)	2336,[375] 2415[381]
Midwall aortic stress (dynes × $10^5/cm^2$)	6·9[283]
Stroke volume (ml)	44,[375] 75–95[58]
Cardiac output (l/min)	5·5,[375] 3·2–3·8,[160] 5·3,[260] 3·79–4·45[165]
Cardiac output (ml/min/kg)	127–143,[51] 115–133[160]
Blood pressure—systemic systolic (torr)	107,[51] 127[160]
Blood pressure—systemic systolic (mm Hg)	115[300]
Blood pressure—systemic diastolic (torr)	74,[51] 91[160]
Blood pressure—systemic diastolic (mm Hg)	102,[300] 83,[162] 104,[357] 111,[375] 109,[300] 83[283]
Blood pressure—pulmonary artery (cm H_2O)	21,[51] 17,[225] 17[375]
Pulmonary artery wedge pressure (cm H_2O)	5·1[225]
Peripheral resistance (cm H_2O × s/ml)	0·22–0·25[51]
Peripheral resistance (dynes × m^2/cm^{-5})	2726,[357] 1840,[375] 2136–2261[165]
Peripheral resistance (mm Hg/l/min)	31·5[160]
Pulmonary blood resistance (cm H_2O/l/min)	3·3[225]

(Table 7, continued)

Pulmonary blood resistance (dynes × s × cm^{-5})	283,[375] 264–270[165]
Oxygen uptake (ml/min)	$29{\cdot}75x + 141{\cdot}1$ (where x = cardiac output (l/min))[58]
Arterial P_{O_2} (torr)	99·5–108·5,[225] 90–98[386]
Arterial P_{O_2} (mm Hg)	89,[375] 86,[260] 89,[171] 77·4,[101] 97,[51] 88,[270] 107,[162] 105[160]
Arterial blood saturation (%)	90·6,[95] 85–87[165]
Arterial P_{CO_2} (torr)	32·4–36·7,[225] 34–35[386]
Arterial P_{CO_2} (mm Hg)	32,[375] 32·1,[171] 31·9,[101] 33·6,[51] 32·1,[270] 40·2,[162] 41·1[160]
Venous P_{CO_2} (mm Hg)	41·3,[125] 37,[163] 62[160]
Arterial pH	7·43–7.45,[225] 7·40,[357] 7·48,[171] 7·40,[101] 7·48,[270] 7·43–7·49,[386] 7·46[162]
Arteriovenous oxygen difference (ml/100 ml)	5·5[101]
Arterial baroreceptor sensitivity (pulse interval/systolic pressure) (msec/mm Hg)	45,[260] 29[201]

can also be produced by extracellular fluid expansion and by administration of ACTH. The latter effect is not dependent on extracellular fluid expansion having a much more rapid onset of action, but is due to mineralocorticoids.[395] A so far unidentified steroid has been suggested.[325] With renal hypertension, vascular sensitivity to angiotensin II is reduced.[235]

3.2 Respiratory System

In addition to the very essential function of gas exchange, sheep use the respiratory system for evaporative cooling. When a sheep becomes hot, it initially pants at a resonate frequency of 150 to 180 breaths/min. This panting is alternated in short periods with short periods of normal respiration and as the heat load for the sheep is increased, there is an

Table 8 Respiratory parameters.

Parameter	Value
Anatomical dead space—bronchi (ml)	101[330]
Anatomical dead space—trachea (ml)	50[330]
Anatomical dead space—naso-buccal (ml)	100[330]
Functional residual capacity (1)	1·35[386]
Tidal volume (ml)	234,[43] 289,[163] 287–455[182]
Tidal volume (ml/kg)	4–9[182]
Respiratory rate (per min)	81,[43] 38,[163] 19,[160] 15–25[182]
Respiratory minute volume (l/min)	19·6,[43] 10·4,[163] 6–7·6[182]
Respiratory minute volume (ml/kg)	100–133[182]
Resting intrapleural pressure (cm H_2O)	−5·6[225]
Alveolar ventilation rate (l/min)	4·1[163]
Dead space ventilation (l/min)	5·8[163]
Pulmonary resistance (cm H_2O/l/sec)	2·8–3·1[386]
Specific pulmonary conductance (per sec/cm H_2O)	0·30–0·32[386]
Static compliance (per l/cm H_2O)	0·13,[225] 0·13–0·15[386]
Volume fraction O_2 in expired air (%)	17·02[182]
Volume fraction CO_2 in expired air (%)	3·02[182]
Oxygen consumption (ml/min)	128,[160] 176–246[182]
Oxygen consumption (ml/min/kg)	3·3–4.2[182]
Alveolar-arterial O_2 gradient (torr)	9·2[225]
Tracheal mucous velocity (mm/min)	17·3[227]

increase in the proportion of time spent panting. While panting, the tidal volume is small (*ca.* 50 ml) but this tidal volume is sufficient for normal levels of oxygen and carbon dioxide to be maintained in the alveoli. This panting is done with minimal extra energy expenditure.[161] When a sheep is exposed to a more intense heat load, the above pattern of panting changes to a different type of panting in which the mouth is open and the breathing is slower and deeper. This second stage panting is a sign that a sheep is distressed, for evaporative heat loss is maximized at the expense of maintenance of alveolar P_{CO_2} (Fig. 9) (but see[187]). The blood pH may rise to 7·75 after an hour of this second stage panting.[163]

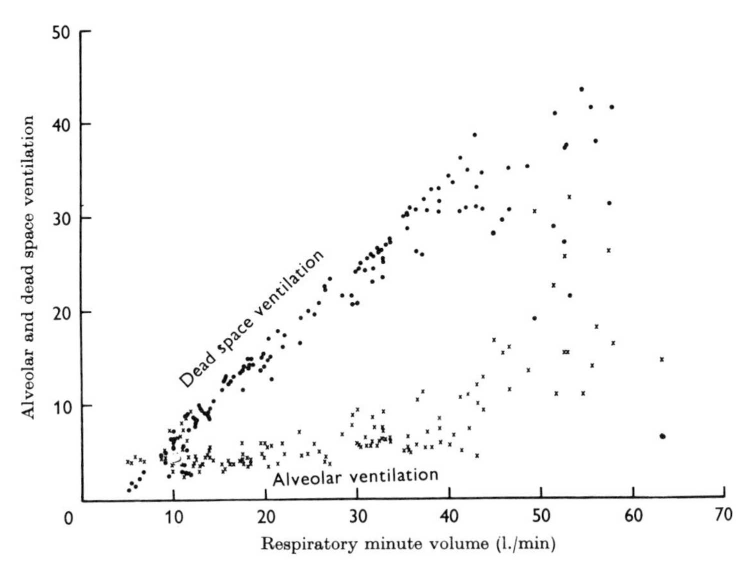

Fig. 9 *Distribution of increased respiratory minute volume into alveolar (x) and dead space (o) ventilation during thermal tachypnoea.*
[From Hales and Webster by permission of Cambridge University Press.]

The range in respiratory rates in Table 8 is probably due to measurements having been made at a variety of laboratory temperatures with the lower values recorded with the sheep at thermoneutral temperatures.

Sheep lungs have a tendency to form atelactaces when the animal is restrained in an unnatural position for a long period. This is particularly apparent in the dependent lung during open chest surgery.

Atelactases formation probably occurs in most large animals when held in unnatural positions. An exception is the human lung which is unusual in that it functions normally for long periods in either the horizontal or vertical position.

Partial pressures of oxygen and carbon dioxide in adult sheep are about 90–100 and 32–40 mm Hg respectively in arterial blood. Published values for oxygen when blood gas analysers are used for measurement are not always adjusted for the slightly higher body temperature of sheep and most of the higher values in Table 7 were obtained with temperature correction. In lambs, the P_AO_2 rises to about 80 mm Hg at 8 weeks and reaches adult values by 20 weeks.[270]

There is some shunting of blood through the lungs which may be the cause of the comparatively low oxygenation of arterial blood (Table 7). Estimates have given this shunting as 4–6%[228] and 7–8%.[101] The degree of shunting may vary in the ewe during the oestrous cycle as exogenous oestrogen can halve the degree of shunting.[95]

Some species (e.g. cattle and man) develop pulmonary hypertension when exposed to high altitudes. In a comparison of several species exposed to an atmosphere equivalent to 4500 m, there was little change in the pulmonary artery pressure in sheep indicating that this species has a tolerance to high altitudes.[363] In fact, sheep can be taken in a decompression chamber in a few minutes to the equivalent altitude of 8500 m (top of mount Everest) with only a transitory bloating of the rumen and need to be taken higher before they lose consciousness.[391] Sheep surprisingly are also able to withstand breathing of 12% carbon dioxide in air for a few days with few apparent ill effects.[195]

In spite of the above, severe hypoxaemia has been shown to cause a doubling of the pulmonary artery pressure.[43,225] The hypoxia also causes regional differences in pulmonary perfusion.[258] Other experimental procedures such as antigen-induced bronchospasm[386] also result in increased pulmonary vascular resistance. Prostaglandin H_2 at a dose rate of 0·25 μm/min/kg will triple the pulmonary vascular resistance without substantially affecting systemic blood pressures and is 100 times more potent than PGE_2 or PGF_{2a}.[51]

3.3 Haematology

The erythrocyte of the sheep is one of the smallest found in mammals and consequently is present in large numbers. The small size can cause difficulties in counting in some models of particle counters set for human blood samples with such counters recording somewhat lower values.

The red cell storage capacity of the spleen (p. 34) has an effect on normal values for red cell parameters (Table 9). Blood samples may be taken from a trained sheep accustomed to the blood sampling procedure, or alternatively the act of sampling may cause splenic contraction with resultant higher values in blood. Anaesthesia causes maximum red cell storage in the spleen and hence minimal red cell values. Rarely in papers are the conditions for blood sampling cited. A

Table 9 Cellular components of blood (see also [157]).

	Values from			
	Schalm *et al.*[320]		Blunt[48]	Winter[403]
Parameter	Average	Range	Range	Mean
Erythrocytes ($\times 10^6$/mm^3)	12	9–15	9·5–13·5	10·6
Haemoglobin (g/100 ml)	11·5	9–15	9–13	13·1
Packed cell volume (%)	35	27–45	30–45	38·0
M.C.V. (μ^3)	34	28–40	30–45	36·1
M.C.H. ($\mu\mu$g)	10	8–12	9–11	12·6
M.C.H.C. (g/100 g)	32·5	31–34	34–38	34.9
Red cell diameter(μ)	4·5	3·2–6·0		
Red cell life span (days)		140–150		
Resistance to hypotonic saline				
minimum		0·58–0·76		
maximum		0·40–0·55		
Platelets ($\times 10^5$/mm^3)	4	2·5–7·5		4–5
Leucocytes ($\times 10^3$/mm^3)	8	4–12	4–12	8
Neutrophils ($\times 10^3$/mm^3)	2·4	0·7–6	0·4–6	
(%)		10–50	30	
Lymphocytes ($\times 10^3$/mm^3)	5	2–9	2–9	
(%)		40–75	60–65	
Monocytes ($\times 10^3$/mm^3)	0·2	0–0·75	0·04–0.8	
(%)			1–6	2–8
Eosinophils ($\times 10^3$/mm^3)	0·4	0–1·0	0·04–1·0	
(%)		0–15	12	
Basophils ($\times 10^3$/mm^3)	0·05	0–0·3	0–0.4	
(%)			0–3	Rare

standard protocol for sampling might be to give an intravenous injection of adrenaline (30 μg/kg) 2 min before sampling.[186]

A decrease over several days or weeks in haematocrit or haemaglobin concentration in sheep held in a laboratory is not necessarily due to anaemia but may be due to sheep responding to the blood sampling procedure by less splenic contraction.

As well as an embryonic haemoglobin which is present up to about day 35,[167] sheep have at least six haemoglobins.[48] Haemoglobin F is present in all foetuses. At birth, it comprises at least 70% of the total haemoglobin but it is completely replaced by adult haemoglobins by 40 days after birth. Studies with erythroid colonies grown from bone marrow cells *in vitro* have shown that the switch from γ to β chain production starts at 115–120 days and is virtually complete at birth.[99] When foetal stem cells were transplanted into adult sheep, they produced adult haemoglobin,[414] showing that it is some factor in the foetal environment that induces haemoglobin F production. Haemoglobin F has the normal sheep α chain and differs from haemoglobin A in the β chain and then only by one amino acid.

Haemoglobin D has been found only in Yugoslavia, and in only three sheep from one flock. In spite of this rarity, its amino acid sequence has been elucidated. It is unusual in that the difference lies in the α chain which has one different amino acid residue. Haemoglobin E, only recently discovered, is a β chain variant found in Indian breeds.[205]

The usual adult haemoglobins are A and B. They differ in the β chain by seven residues. Red cells of sheep which are heterozygous for A and B genes have both haemoglobins. Under appropriate conditions, haemoglobin A and AB but not B cells will sickle.[64]

Probably the most interesting sheep haemoglobin is C for which there is no precedent in the animal kingdom.[364] It differs in that it is four residues shorter than other haemoglobins and has 17 residues different to the β chain of haemoglobin A. Is is found only in haemoglobin A or AB sheep and then only after a severe anaemic stress has been imposed. It appears in blood about one week after the imposition of the anaemic stimulus and may comprise up to 20% of haemoglobin. Its production depends on both the severity of the anaemia and the rate at which the anaemia is produced. Haemoglobin C seems to offer no physiological advantage during anaemia.

A measure of the oxygen unloading capacity of a haemoglobin is the

P_{50} or partial pressure of oxygen at which the haemoglobin is 50% saturated with oxygen. Haemoglobin B has a lower oxygen affinity than haemoglobin A (Table 10) but this does not seem to be a handicap to haemoglobin B sheep. The oxygen saturation curve for haemoglobin C is identical to that for haemoglobin A[75] but the oxygen affinity of haemoglobin C is much more sensitive to the P_{CO_2} than that of haemoglobin A.

Table 10 Oxygen affinity constants (P_{50}) of sheep haemoglobins (modified from[272]).

Haemoglobin	Naughton *et al.*[279] (pH 7·4, 38°C)	Huisman and Kitchens[198] (pH 7·35, 37°C)
A	27·7 mm Hg	29·5 mm Hg
B	37·8	43·0
AB	32·7	
C (86% C − A 14%)		29.1
Newborn (82% F − 18% B)		25·7
F	16·0	

There are only low levels of pentose phosphate enzymes in sheep erythrocytes but they are adequate to maintain a satisfactory GSH: GSSG ratio and the cells are as resistant to oxidative drugs as human erythrocytes.[352]

Unlike human haemoglobins, there is very little binding of diphosphoglyceride to sheep haemoglobins.[27]

Sheep red cells may have either a high or a low potassium concentration (p. 53). At birth, all red blood cells have a high intracellular concentration of potassium but a few weeks after birth the soluble antigen *L* or the soluble antigen *M* appears in blood and becomes adsorbed onto the red cell surface to become one of the red cell antigens. The *L* antigen partially inhibits the activity of the sodium-potassium pump (possibly by inhibition of Na-K activated ATP-ase in the membrane) and the intracellular potassium concentration drops as a result.[376]

Methodological differences can influence the derivation of values for lifespans of red cells but the concensus of opinion is now that sheep red cells have a life span of about 140 days and that there are no differences between cells with haemoglobins A, B or C. As foetal red cells are lost relatively rapidly after birth, they must have a shorter life span. Low glutathione red cells of the Merino type (p. 53) have a normal lifespan but those of the Finnish Landrace type have a shortened lifespan.[365]

An unusual feature of sheep leucocytes is the low ratio of neutrophils to lymphocytes. This does not imply an inability to combat disease as the neutrophil count increases with bacterial infections. Most of the lymphocytes are small (8–10 μm) lymphocytes.

An area of importance lies in the closeness of sheep coagulation parameters to those of humans as these determine the value of the sheep as a model for development of intravascular devices for eventual use in man. The most thorough comparison which has been done is that by Gajewski and Povar[145] (Table 11). The most significant apparent difference is the plasminogen level which in the data of Gajewski and Povar is only 5% of that in human blood. This low level

Table 11 Comparison between sheep and humans of blood coagulation parameters (see also [133]).

Parameters	Sheep mean ± S.D.	Human range
Whole blood clotting time (glass) (min)	10 ± 3	5–12
Whole blood clotting time (plastic) (min)	31 ± 17	20–50
Activated partial thromboplastin time (s)	32 ± 10	28–44
Prothrombin time (s)	15 ± 1	13–15
Thrombin time (% of human mean[a])	108 ± 18	100
Factor II (% of human mean[a])	21 ± 11	60–140
Factor V (% of human mean[a])	417 ± 227	60–140
Factors VII and X (% of human mean[a])	51 ± 34	60–140
Factor VIII (% of human mean[a])	809 ± 408	50–200
Factor IX (% of human mean[a])	211 ± 91	60–140
Factor X (% of human mean[a])	32 ± 20	60–140
Factor XI (% of human mean[a])	37 ± 26	60–140
Platelet count ($\times 10^3/mm^3$)	457 ± 121	200–400
Platelet adhesiveness (% adhered)	80 ± 14	18–60
Plasminogen level (units/ml)	86 ± 12	2000–4000
Antiplasmin level (units/ml)	158 ±21	80–150
Fibrinogen level (mg/ml)	292 ± 81	200–400

[a] These values were calculated from a dilution curve of normal human values. Values were obtained from 25 sheep and an unstated number of humans.[145]

may however be an artifact as sheep plasminogen is not activated by streptokinase for the reason that the initial small amounts of plasminogen that are formed inactivate the streptokinase.[287] Urokinase however, is not broken down by plasminogen and hence will activate plasminogen. Other differences in the data in Table 11 are small and it is doubtful if they are physiologically significant. The differences in values for Factor VIII in Table 11 were found by Fantl and Ward[133] but not by Lopacuik *et al.*[243]

Some differences in coagulation parameters may be subtle as one study has shown that low molecular weight heparins are more active in sheep than in human plasma even though sheep plasma in general required more heparin for anticoagulation than human plasma.[332]

The dog and the rabbit have been used widely as test animals for biocompatability of plastics in contact with blood. Gabrowski *et al.*[152] compared platelet function of several species including the sheep, dog, rabbit and human. Sheep, non-human primate and pig platelets behaved similarly to human platelets in their adhesion to a test plastic whereas rabbit and dog platelets were much more adhesive. However there still are differences in platelet function as sheep (and Rhesus monkey and pig) platelets are less sensitive to ADP and do not respond to adrenaline by aggregation and release of serotonin.[2] Also sheep platelets require about ten times more prostacyclin to inhibit aggregation than do human platelets.[116]

The above data suggest that the sheep is reasonably well suited for testing of plastics and artificial organs for blood compatability.

The composition of plasma is indicated in Tables 12 to 16. Most of the values are similar to other laboratory animals. As well as the normal amino acids, there are large amounts of citrulline and unusual amino acids such as 1-methyl and 3-methyl histidine and α-amino butyric acid which are associated with products of ruminant digestion. A small proportion of the blood amino acids may be in the *D*-form, these being derived from microbial cell walls. *D*-amino acid oxidases are widely distributed in mammals and so levels will be low. Appreciable amounts of amino acids are also transported in red blood cells and for some amino acids the intracellular concentration can be higher than in plasma.[179]

Plasma proteins are in general as expected. However haptoglobin is not normally detectable in sheep but several haptoglobins can be found on the induction of stress.[203]

Table 12 Inorganic constituents and osmolality of plasma.

Element	Concentration
Sodium (mM or mEq/l)	147,[125] 163,[29] 145[340]
Potassium (mM or mEq/l)	4·85,[125] 4·6,[29] 5·2[340]
Calcium (mM/l)	2·46,[29] 2·3–2·7,[340] 2·3[340]
Calcium (mg/l)	71–90[338]
Magnesium (mM/l)	0·84[29]
Chloride (mM/l)	112,[29] 107,[125] 104[340]
Phosphorus (mM/l)	2·07[29]
Phosphorus (mg/l)	40–56,[338] 47[340]
Copper (mM/l)	160[340]
Copper (mg/l)	0·88[351]
Cobalt (μg/l)	0·25–3.0[340]
Iron (μM/l)[a]	27–35[340]
Zinc (μM/l)	15[340]
Fluoride (μM/l)	10[340]
Osmolality (mOsmol/l)	293,[125] 304[29]

[a] Transferrin saturation = 38%

Table 13 Amino acids in plasma.

Amino acid	Morris et al.[275] (Fed) mM/l	Slater and Mellor[336] (Fed) mM/l	Bergman and Heitmann[37] (Fed) mM/l	Bergman and Heitmann[37] (Fasted) mM/l	Tao et al.[348] (Fed) mg/100 ml
Alanine		124	104	94	2·24
Arginine	71	145	144	81	2·01
Asparagine		36			
Aspartic acid					0·39
α-amino butyric acid	23				
Citrulline	143	187	180	65	
Cystine					0·14
Glutamic acid	103	125	55	56	2·39
Glutamine		208	219	173	
Glycine	550	826	348	559	4·07
Histidine	50	37			0·21
1 and 3 methyl histidine	128	58			
Leucine	128	67	148	126	1·72

Table 13 Amino acids in plasma.—continued

Amino acid	Morris *et al.*[275] (Fed) mM/l	Slater and Mellor[336] (Fed) mM/l	Bergman and Heitmann[37] (Fed) mM/l	Bergman and Heitmann[37] (Fasted) mM/l	Tao *et al.*[348] (Fed) mg/100 ml
Isoleucine	84	79	101	89	1·41
Lysine		131			0·85
Methionine		14			0·30
Ornithine		71	100	65	
Phenylalanine	47	35	52	38	0·85
Proline					1·53
Hydroxyproline					0·43
Serine	75	95	49	28	0·79
Taurine	344	84			
Threonine		52	125	48	1·48
Tyrosine		38	58	25	1·51
Valine		114	268	184	2·38

Table 14 Plasma lipids (from [232]).

Fraction	Amount (mg/100 ml)	Percentage: Cholesterol	Cholesterol esters	Triglyceride	Phospholipid
Very high density lipoproteins	*n.s.*[a]	1·4	0·6	0	17·0
High density lipoproteins	261·8	52·3	72·7	4·4	65·8
Low density lipoproteins	67·4	42·9	26·1	23·0	14·7
Very low density lipoproteins and chylomicrons	10·7 } 4·3 }	3·5	0·5	72·6	2·0

[a] *n.s.* not stated.

Table 15 Organic constituents in plasma.

Substance	Concentration
Lactate (mM/l)	1382,[184] 950–1240,[239] 1300–1600[210]
Pyruvate (mM/l)	73,[184] 50–110[239]
3-hydroxy butyrate (mM/l)	240,[184] 0–250[239]
Acetoacetate (mM/l)	29,[184] 10–86,[239] 160–290[210]
2-oxoglutarate (mM/l)	20,[184] 10–26[239]
Urea (mM/l)	5819[184]
Glucose (mM/l)	4000,[184] 2710–3750[210]
Citrate (mM/l)	161[184]
Succinate (mM/l)	30–70[239]
Oxaloacetate (mM/l)	12[239]
Formate (mM/l)	28–120[239]
Acetate (mM/l)	400–1500[239]
Creatinine (mg/l)	0·70[125]
β-hydroxybutyrate (mM/l)	160–290[210]
Free fatty acids (mM/l)	610–1110[210]
Albumin (g/dl)	4·1[340]
Globulin (g/dl)	2·8[340]

Table 16 Serum enzyme levels.

Enzyme	Activity (μmol/l/min)
Alkaline phosphatase	225–276,[212] 64,[340] 21[221]
Arginase	0·37[212]
Creatine phosphokinase	83,[340] 50[221]
Fructose 1-phosphate aldolase	3·8–5·4[413]
Fructose 1,6-diphosphate aldolase	16–27,[413] 23[417]
Lactic dehydrogenase	664–845,[413] 311,[340] 313[221]
Malic dehydrogenase	340,[417] 826–2113[413]
Serum glutamic oxaloacetic transaminase	36–62,[417] 70–122,[413] 24[221]
Serum glutamic pyruvic transaminase	17·4,[212] 8–9[417]
TPN-linked isocitrate dehydrogenase	255–276[212]

(See [4,172,211,352] for blood enzyme levels).

3.4 Immunology

Most immunologists have been involved with sheep in their research as sheep red cells are the standard small particle used in many immunological tests such as the haemagglutination tests and rosette formation tests for T-lymphocytes. (See also [98].)

Although there are several antigen systems found on the red cell of the sheep (p. 52), the only antigen system for which natural antibodies are present is the *R/O* system. Anti-R antibodies show seasonal variations in titres, reaching highest levels in summer and lowest in winter.[366] Proportions of Ig G antibodies to different antigens vary during primary and secondary responses.[262] The sheep appears to have smaller responses (i.e. produces lower titres) to both primary and secondary challenges than does the rabbit.[214] Another difference from the rabbit and also from man is that the predominant immunoglobin is Ig G_1 rather than Ig G_2. Typical values for serum immunoglobins are:

Ig G_1	Ig G_2	Ig M	Ig A	
9·4	1·8	1·75	0·11	mg/ml[291]
18–19	5·8–6·3	1·2–2·0	0·17–0·31	mg/ml[65]

In spite of these differences, the sheep has been a useful animal for antibody production.[49,345]

The lamb becomes immunologically competent well before birth.[87] Lymphocytes appear at 32 days gestation, monocytes at 63 days, eosinophils at 112 days, Ig G at 56 days and Ig M at 77 days.[317] Challenging foetuses with several antigens showed development of antibodies to all but one at around 64 to 82 days.[131,133] Thymectomy done at 66–88 days together with or without splenectomy reduced but did not abolish immune responses.[86,291]

Transfer of maternal antibodies to the lamb occurs, as in most other animals, only from the colostrum and the antibodies (Ig G)[135] can be absorbed from the intestine for only about 36 h after birth. Over 70% of the immunoglobin in colostrum is Ig G_1.[65] Up to 10 g of gamma globulin is absorbed and this can double the plasma protein level in the lamb's blood.[335] There are suggestions that absorption of antibodies can be variable and that low absorption by individual lambs can enhance the susceptibility to disease.[318]

Probably the most interesting immunological research on sheep has been on the antigenic stimulation of single lymph nodes which have had the efferent ducts cannulated for complete collection of lymph (see[174]). A typical lymph node produces 5–10 ml of lymph per hour containing 5000 to 20 000 lymphocytes per mm^3. A node weighing 1 g will discharge 3×10^7 lymphocytes/h of which most will have been filtered from blood perfusing the node.[164] After antigenic stimulation of a node, there is a fall in the output of lymphocytes lasting for a few hours but the lymph flow rate does not change. Then for about 48 h lymphocytes increase in numbers by 2–3 times. Starting at 50 to 70 h and with a peak at 80 to 120 h, large pyroninophillic immunoblasts appear and these comprise 40% of the cells in lymph. These immunoblasts have been shown *in vitro* to be capable of conversion to plasma cells. The cellular flow in the lymph from the stimulated node returns to normal by 200 h.[164]

Transfer of immunoblasts to a chimeric twin causes antibody production in the recipient in 2–3 days.[164] If all of the cells draining a node are diverted from the animal for a few days, the node which was stimulated will have retained its immunological memory (i.e. it will give a secondary immune response when stimulated with the same antigen) but all other nodes will act as if they had had no previous encounter with the antigen.[339] Transfer of lymph draining a stimulated node to another node will cause that node to produce antibodies. These experiments show that antigenic stimulation of a lymph node results in the production of cells which colonize other lymph nodes and thereby induce antibody production. The recent technique for creating identical twin lambs (p. 66) may stimulate further research involving transfer of immunological cells.

If instead of injecting an antigen into tissue drained by a node, a non-pathogenic virus is injected, there is no initial fall in cell output from the node but instead an immediate increase of 5–10 fold in lymphocytes.[292]

In contrast to peripheral lymph nodes, intestinal lymph contains a mixture of Ig G_1- and Ig A-containing blast cells. Antigenic stimulation of an intestinal lymph node via a cannulated afferent duct produces mainly Ig G_1-containing cells while injecting the antigen into the intestinal wall and collecting lymph from an afferent lymphatic shows the stimulation of Ig A-containing cells from the lamnia propria and Peyer's patches of the intestine.[31]

Cutaneous antigen–antibody reactions in the sheep are of the reversed passive Arthus type with the vascular permeability changes, unlike in the rat, being due to histamine.[377] Anaphylactic responses are characterized by profound systemic arterial hypotension and systemic venous and pulmonary hypertension accompanied by peripheral pooling of blood in the pulmonary and mesenteric vascular beds.[130]

Compared with the research that has been done on sheep red cell antigens, little is known about histocompatability antigens. Several lymphocyte antigens have been defined of which nine are produced by multiple alleles at two closely linked loci.[21,139,269,323] An interesting observation is that many multiparous ewes have antibodies which react with these antigens whereas neither rams nor lambs have such antibodies. The suggestion is that immunization of the ewe with lamb histocompatability antigens occurs during pregnancy. This is known to occur in women.[140] As there is no placental transfer of antibodies, this maternal immunization would not harm the lamb *in utero*.

In sheep as in cattle, horses and goats, complement components C_2, C_3, C_4 and C_5 appear to be undetectable or only present in low titres but their detection may have been made difficult by species specificity.[19] Because sheep red cells are used in the haemolytic assay usually used for assays of complement, conditions for assay of sheep complement need to be changed slightly.[20] New born lambs have about half of the level of complement of their mothers.[305]

3.5 Genetics

Considerable research has been done on the genetics of sheep. The most comprehensive account, *Quantitative Genetics of Sheep Breeding*[373] deals primarily with multigene factors and how they can be manipulated for increasing production from sheep. Other accounts with considerable details are given by Ryder and Stephenson[314] and Tucker.[366]

The normal chromosome number is $2n = 54$. Three different chromosomal translocations (Robertsonian translocations) have been studied and these, unlike in some other species, have no effect on fertility. In fact, a ram and a ewe which each had the three of these translocations were mated and produced a lamb which, like the parents, had the chromosomal number reduced to $2n = 48$.[61]

Several genetically controlled red cell features have been found (Table 17). Haemoglobin type was discussed earlier (p. 42). Eight systems of red cell antigens have been identified. The *R* or *R–O* system is the most important as naturally occurring haemolytic anti-*O* or anti-*R* antibodies may be present.[366] The red cells do not contain these antigens at birth but like *ABO* antigens in man acquire them from the plasma soon after birth. The *R* antigens are also present in saliva, seminal plasma and other secretions. A complication of the *R* system is that the *I* gene controls its expression.[364] The *A*, *B*, *C* and *M* systems are also haemolysins whereas the *D* system is agglutinating but naturally occurring antibodies are never found to these systems. The *M/L* system has an association with the gene for red cell potassium (see below).

Table 17 Genetic markers in erythrocytes (modified from [366, 367]).

Factor	Locus symbol	Alleles	Number of phenotypes
Blood group antigens[366]	*R*	*R* r^o	3
	I	*I i*	
	A	*a b* —	4
	C	*a ab b* —	4
	M	*a b ac c*	6
	D	*a* —	2
	B	*a ab abc* — etc.	Many
	X–Z	*X* X^z	3
	Hel	*Hel Hel*	3
Haemoglobin[366]	*Hb*	*A B (C, D)*	3
Carbonic anhydrase[366]	*CA*	*F S*	2
"X" protein[366]	"*X*"	*X x*	2
Purine nucleoside phosphorylase[371]	*NP*	*H L*	2
Potassium level[366]	*Ke*	*L H*	2
Lysine[366]	*Ly*	*A a*	2
Glutathione[369]	*Tr*	*H L*	2
	GSH	*H L*	2
Diaphorase[368]	*Dia*	*F S*	3
Arginase[408]	*Arg*	*A a*	2

(See also [261]).

At least seven blood proteins have genetic alleles (Table 18). Most of the allelic systems are simple but the transferins have multiple alleles and many phenotypes.

Table 18 Genetic markers in plasma (modified from [366, 367]).

Protein or marker	Locus symbol	Alleles	Number of phenotypes
Albumin[366]	*Al*	*D F S T V W*	10
Prealbumin[366]	*Pr*	*F S O*	6
Prealbumin[366]	*X*	*F M S*	6
Transferin[366]	*Tf*	*I A G B* and others	Many
Esterase[366]	*Es*	*a b c*	6
β Lipoprotein[366]	*Lp*	1 —	3
α_2 Macroglobulin[366]	*ap*	1 2	3
Immunoglobulin[366]	*Im (1)*	$\gamma^{1a}\gamma^{1b}$	
Immunoglobulin[366]	*Im (2)*	γ^{2}	
Haptoglobin[32]	*Hp*	*a b c*	
Antigen[200]	A_1	A_1 A_2 A_0	3
Antigen[33]	B_1		

(See also [261]).

Most mammalian cells have a high intracellular content of potassium and a low content of sodium which is the reverse of the extracellular electrolyte situation. Some species however, have a low erythrocyte potassium content which is balanced by a high content of sodium. In sheep erythrocytes, either state can be found with the potassium to sodium ratio being either about 80:20 or 20:80.[3] These are referred to as HK and LK sheep respectively. The reason for the difference lies in the potassium pump mechanism as HK cells have a greater ouabain-sensitive potassium influx, a greater number of ouabain binding sites per cell and greater sodium-activated ATPase.[367] The difference in electrolyte content has no apparent effect on the physiological function of the cells.

Two sets of genes control red cell glutathione level. Glutathione is thought to protect cells from oxidative damage to the haemoglobin. In one of the systems, found so far only in the Finnish Landrace breed, the low glutathione level is associated with high intracellular levels of lysine and ornithine due to a defective transport system for certain amino acids and the red cell lifespan is shortened. The other system found in several breeds including the Merino is associated with an impaired activity of the enzyme γ-glutamyl cysteine synthetase and red cell lifespan is not affected.[369]

Coat colour is complicated and is controlled by several gene systems. In various breeds, there is a dominant gene for black, a recessive gene for black, a dominant gene for brown and a recessive gene for brown. For each of these genes, the second allele is white. Other genes control the distribution and pattern of colour and account for amongst other features the black or white faces of different breeds. Still more genes control the characteristics of the fleece[314] but often they are multigene systems and are not capable of simple description.

Two gene systems control horns. These will cause both sexes to be horned, both sexes to be hornless (polled) or the rams to be horned and the ewes polled. Breeds tend to be homogenous for these genes.

Freemartins (abnormal development of the female genitalia of the female member of non-identical twins) are common in cattle but rare in sheep. Fusion of the foetal membranes can be demonstrated when they occur. Unlike cattle, there is extreme masculinization of the genitalia of both twins[397] and they cannot easily be identified as freemartins.

Chimeras occur naturally and can be formed experimentally.[364,370] Proportions of cells lines can be stable or can oscillate widely.[100]

The occurrence of various forms of chromosomal abnormalities has been noted and the frequency of such abnormalities including trisomy might be about 6%.[241]

3.6 Nervous System

Comparatively little research has been done on the nervous system. Stereotaxic atlases have been published for the brain[307,393,253] and visual cortex.[83] Several papers deal with projections to the brain and pathways in the spinal cord. They include mechanosensory projections[206] and retinal projections to the cerebral cortex.[215]

Some experiments have involved lesioning or stimulation by electrodes of parts of the hypothalamus. Salt appetite could be electrically stimulated in some but not all sheep.[252] It is more difficult in sheep than in monogastric animals to produce hyperphagia[349] but diabetes could reliably be produced by lesioning.[59] Lesioning has also indicated the presence of both stimulatory and inhibitory areas controlling mammogenic and lactogenic processes.[112]

In wethers at least, negative feedback for LH secretion appears to be through oestradiol rather than testosterone or progesterone concentrations.[120] Androstenedione may also play a role as active immunization of ewes against this steroid increases ovulation rate and oestrogen level in the periovulatory period.[319]

Control of food intake presents an interesting problem. Gold thioglucose does not produce hypothalamic changes even when given in massive doses.[16] Several drugs when given intraventricularly, systemically or in feed will either initiate or increase feeding.[15] A hormone which has recently been suggested as being important in the control of feeding in ruminants is cholecystokinin.[107] Exchange of blood between hungry and satiated sheep caused the satiated sheep to begin eating soon after while reducing food intake in the hungry sheep. Cerebrospinal fluid transferred from the lateral ventricle of hungry sheep to that of satiated sheep also elicited feeding in the recipient.[14]

In grazing sheep or sheep given roughage diets, the two main factors controlling food intake are palatability of the food and the physical capacity of the rumen.[11] Less digestible foods tend to be less palatable but also are retained in the rumen for longer and this longer retention time with the volume factor also limit food intake.

Initiation and cessation of feeding in sheep fed largely on concentrates appears to be due to signals other than "fill" but is more complex than in monogastric animals.[14,15] One factor inducing cessation of eating may be the fluid shift into the rumen and the associated decrease in blood volume (p. 85).

Apomorphine causes "internal vomiting" whereby digesta is regurgitated from the abomasum (true stomach) to the rumen. It has been suggested from this that there may be two vomiting centres in the brain of the sheep.[121]

There are temperature receptors in the hypothalamus, in skin and in the abdomen, possibly in the wall of the rumen and the small intestine.[303]

The new born lamb and the ewe at parturition are not responsive to injections of pyrogens.[92] It is probable that prior sensitization is required for typical fever response in lambs. Parturient ewes would have been sensitized and the explanation that has been advanced for their lack of fever response is a response to vasopressin released at parturition.[92]

The "diving" reflex in which respiration is suppressed when the nostrils are covered with water is present in sheep.[169]

Administration of tryptophan intravenously will increase the level of this amino acid and its metabolites in the brain. One metabolite, serotonin, has been shown to increase in the mid-brain and hypothalamus.[13]

The composition of cerebrospinal fluid is given in Table 23. Cannulae in the third ventricle and the cysterna magna allow perfusion of artificial cerebrospinal fluid through the brain.[276] In such experiments, lowering the calcium level below 2 mg/100 ml or magnesium below 0·6 mg/100 ml produced hyperpnoea and continuous muscle tremors and episodes of tetany respectively.[7] It has been suggested that an increase in osmolality of the c.s.f. can lead to arterial hypertension.[155]

The peripheral nervous system presents few peculiarities. The distribution of the dermatomes is given by Kirk.[223] One peculiarity is the distribution of the cardiac accelerator nerves[381] which enter the thoracic sympathetic chain through spinal roots T_1–T_6 on the right side and T_2–T_5 on the left side and may ascend or descend a considerable distance (to T_{10}) before returning to the heart. They may be absent on the left side.

Finally, neurological examination of sheep has been reported to be disappointing and may produce such unsatisfactory results that a veterinary clinician "would be very severely hampered in making a proper diagnosis in many kinds of pathology of the central nervous system".[188]

3.7 Endocrinology

Cited values for hormone concentrations that have been measured in sheep are given in Tables 19–21 (see also [136,137]). These have been obtained mainly with radioimmunoassay or competitive protein binding assays which have greater sensitivity than the types of hormone assays used several years ago. The greater sensitivity has permitted smaller blood samples to be taken and hence has facilitated

Table 19 Plasma concentrations of some hormones.

Hormone	Concentration	Comments
Anterior pituitary gland		
Growth hormone	1·9–2·1 ng/ml[105]	Ewes
	10·2 ng/ml[103]	Rams
	3·6 ng/ml[103]	Wethers
ACTH	21–63 pg/ml[209]	
TSH	1·9–2·6 ng/ml[104]	
LH, FSH and Prolactin—see Table 21		
Melatonin	140 pg/ml[310]	Day time
	297 pg/ml[310]	Night time
Posterior pituitary gland		
Vasopressin	5 μU/ml[208]	Arterial blood
	20 μU/ml[208]	After haemorrhage
Thyroid and parathyroid glands		
T_3	0·65 ng/ml[78]	
rT^3	0·61 ng/ml[78]	
Thyroxine (T_4)	0·057 μg/ml[78]	
Calcitonin	5·3 ng/ml[147]	Pregnant ewes
	<0·5 ng/ml[147]	Male lambs
Parathormone	0·53–0·67 ng/ml[147]	
Gut and pancreas		
Glucagon	20 μU/ml[54]	
Insulin	260 pg/ml[54]	
	0·8 ng/nl[359]	Fed
	0·3 ng/ml[359]	Fasted
Other glands		
Cortisol	12 pg/ml[285]	
	4·0–6·3 ng/ml[189]	Rams
Aldosterone	<1–2 ng/100 ml[276]	
Renin	1·2–1·3 units[70]	Non-pregnant ewes
	0·36 units[70]	Pregnant ewes
Renin substrate	99 ng/ml[70]	Non-pregnant ewes
	392 ng/ml[70]	Pregnant ewes
Angiotensin II	15–20 pg/ml[245]	

frequent sampling (as often as every 10 min for 24 h)[144] which in turn has produced results which have changed some endocrinological concepts. Frequent blood sampling in sheep has shown that many of the hormones (LH and testosterone[216,302], LH and oestradiol,[17] progesterone,[255] cortisol,[142,144,256] prolactin and TSH,[104] relaxin,[76] growth hormone[103]) show episodic secretion patterns (ultradian rhythms) and as a consequence the fluctuations in concentrations can be much greater than the mean of a number of samples. Consequently the value of taking only one or two samples may be limited. For realistic estimations of hormone status in future research, either several samples should be taken at timed intervals or else a device which takes a continuous slow blood sample over a sufficiently long period to even out fluctuations in hormone levels should be used.

The peaks are sufficiently frequent to allow a significant base line to be maintained above zero for some hormones (e.g. cortisol, progesterone (Fig. 10)). For others (e.g. testosterone in rams), the peaks are less frequent and so the base line is near zero. With at least testosterone

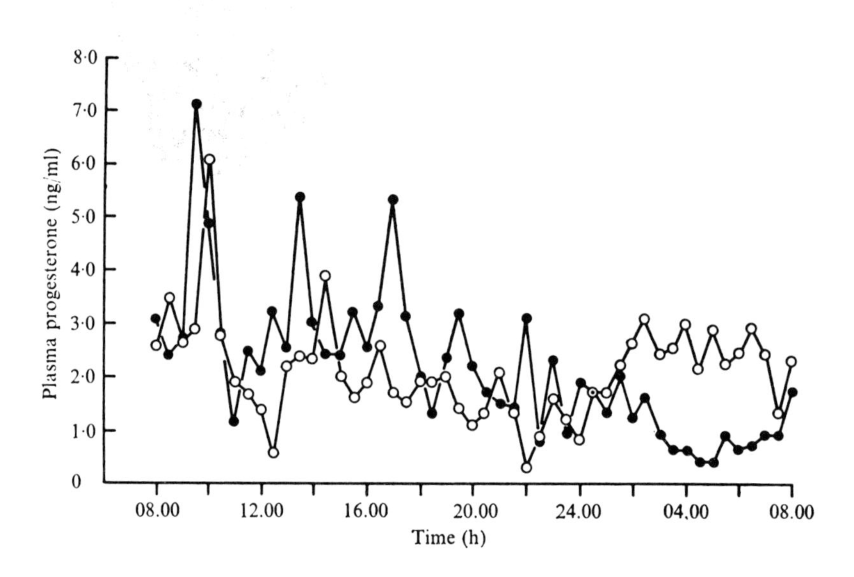

Fig. 10 *Variation in plasma progesterone concentration in two sheep during a 24 h period.*
[From McNatty *et al.*[255] by permission from the *Journal of Endocrinology.*]

in rams, the critical factor may be the frequency of the peaks rather than the mean level or the average height of peaks.[237] This may also apply to hormones in the ewe as Baird[17] showed that smaller but more frequent LH pulses at oestrus as compared with the luteal phase caused greater oestradiol secretion.

These episodic secretion patterns are also known for many hormones in man. Restriction in the number of blood samples that can be taken make their identification in small laboratory animals more difficult than in sheep.

Some fluctuations in hormone levels can be ascribed to changes in metabolic clearance rates as many hormones, especially steroid hormones, are largely metabolized in the liver. As an example, 48% of the metabolic clearance rate of cortisol occurs in the liver.[284] Fasting and anaesthesia reduce gastro-intestinal blood flow and hence can increase hormone levels in systemic blood.[156]

Diurnal (circadian) fluctuations are also present in many hormone profiles in sheep (e.g. progesterone,[255] melatonin,[265] cortisol,[144,187,316] but not LH[173]). Hence sampling should be done where possible at standard times. Some of the diurnal fluctuations may be due to the effects of once daily feeding regimes and may be absent with more frequent feeding. Feeding is most likely to have an effect on hormones which are metabolized in the liver (e.g. steroids, but not insulin[55]) as suggested above. However the rise in melatonin seen at night is a direct consequence of a change in illumination as concentrations changed dramatically when lights were switched on or off.[310]

Details of clearance rates and other kinetic data are given in Table 20.

It must always be borne in mind that the act of sampling may influence hormone secretion. This is likely to happen with cortisol unless animals are accustomed to the sampling procedure (see p. 129). It may have happened with growth hormone and prolactin in the experiments of Malvern and Mollett.[259] It does not happen with all hormones as for example secretion of LH was not affected by venepuncture, vascular cannulation, starvation or running sheep around a yard for 5 min before sampling.[173]

Several hormones are carried on blood proteins. Just over half of the thyroxine is carried on thyroid binding globulin and another third on what appears to be another specialized thyroxine binding blood protein.[346] Cortisol is carried on transcortin (59%) and albumin (19%) while about 22% is free in plasma.[290]

Table 20 Half life, production rate and metabolic clearance rate of some hormones.

Hormone	Half life	Production rate	Metabolic clearance rate
Growth hormone[359]	9–13 min	28 μg/h	8–14 l/h
ACTH[209]	0·8–1·3 min	46–107 μg/kg/h	34 ml/kg/min
Prolactin anoestrus ewes[102]		130 μg/h	3·8 l/h
Prolactin lactating ewes[102]		2520 μg/h	8·5 l/h
Prolactin pregnant ewes[102]		262 μg/h	6·8 l/h
LH[5]	24 min	0·5–84 μg/h	2·6–2·9 l/h
FSH[5]	120 min	138–426 μg/h	1·2–1·7 l/h
Cortisol[285]		600 μg/h	51 l/h
Oestradiol-17β[74]		0·4–6·2 μg/h	147 l/h
Oestrone[73]		8·5–14·4 μg/h	206 l/h
Ovine placental lactogen[168]	29 min		
Progesterone[30]		36–4330 μg/h	209–300 l/h
TSHRH[224]	20, 44 min		54 l/h
T_3[78]		1·3 μg/h	2·1 l/h
rT_3[78]		1·7 μg/h	3·4 l/h
T_4[78]		7 μg/h	0·12 l/h
Insulin[55]	12–13 min	0·43 *U*/h	

A congenital goitre has been found in Merino sheep.[111] In this goitre, thyroglobulin in the thyroid colloid is replaced by a thyroglobulin-like iodoprotein.

As in man and the rat, the stores of LH in the pituitary gland appear to be in two pools of which the first is released by an intravenous injection of a small amount of LH releasing hormone and the second after a similar injection given 1 to 3 h later.[96] The absolute amounts and the relative pool sizes vary during the oestrous cycle of the ewe.[344] There may be a direct feedback loop for pituitary hormones to the brain as saggital sinus blood can contain much more ACTH than carotid blood.[36]

Prostaglandins are of rapidly increasing importance in research. PGF_2-metabolizing enzymes show differences in location in the uterus of pregnant sheep compared with the pregnant woman. There are also differences in metabolic pathways.[220] An enzyme in sheep red cells but not in sheep plasma or in blood from several other species used as laboratory animals will convert PGE_2 to PGF_{2a}. It has optima of 37°C and pH 7.0 and is destroyed by incubation of blood at 55°C.[183] To avoid its action, blood should be centrifuged immediately if prostaglandins are to be measured.

3.8 Reproduction

Although rams will mate and are fertile at any time of the year, they have higher levels of reproductive hormones during winter when ewes are having regular oestrous. The average ejaculate from a ram is about 1 ml and contains $3{\cdot}6 \times 10^6$ sperm. A ram can ejaculate 20 to 40 times per day without a significant decrease in sperm number in each ejaculate.[266] However extensive "work" can cause a decline in semen characteristics; in one paper from $1{\cdot}6 \times 10^6$ (85% live) to $0{\cdot}4 \times 10^6$ (63% live) per ejaculate.[334] Usually a ewe in oestrus will be served by a ram or rams several times but in each ejaculate there are sufficient numbers of sperm to inseminate 20 or more ewes by artificial insemination.[266] Although up to 5% of rams (1 for 20 ewes) are used in flocks, as low as 1% may be adequate. A common ratio of rams to ewes is 2%. Even when ewes are synchronized so as to come on oestrus within 3 to 5 days of each other, 5% of rams are considered sufficient,[229] probably because after a brief busy period rams will only be lightly "worked".

Ewes are seasonally polyoestrus which means that (unless mated) they will regularly exhibit oestrous behaviour ("cycle") for several months of the year. The seasonality is controlled by a shortening of day length so that oestrus occurs in late autumn and winter. Breeds vary in their seasonality and some ewes of the Merino and Dorset Horn breeds may show oestrus at any time of the year although the highest proportion cycle in late autumn. Other breeds may have a very restricted breeding season and this is influenced in general by the latitude from where they originated and that where they are kept. Ewes transferred from one hemisphere to the other adopt the seasonality of the new hemisphere after a period of about a year during which time the pattern of oestrus can be irregular.

The seasonal anoestrus may be due to prolactin[277] as shortening of day length produces no change in FSH or LH but a decrease in prolactin level.[385]

The day on which a ewe first shows oestrus is conventionally termed day 0 of the cycle. Cycles last for about 17 days in Merino and

16 days in most British breeds with a standard deviation of 1 day. Oestrus may last for 36 h but can be as short as 8 h in young ewes and in ewes near the start or end of the breeding season.

The number of ovulations is influenced by nutrition but is largely genetically controlled.[40,226] Active immunization of ewes against androstenedione increases the number of ovulations but does not increase the number of lambs born.[264]

Both prolactin and LH are required for luteal function.[109]

Details of levels of reproductive hormone are shown in Table 21 and Fig. 11. The active oestrogen at oestrus is oestradiol 17β. Peaks of this

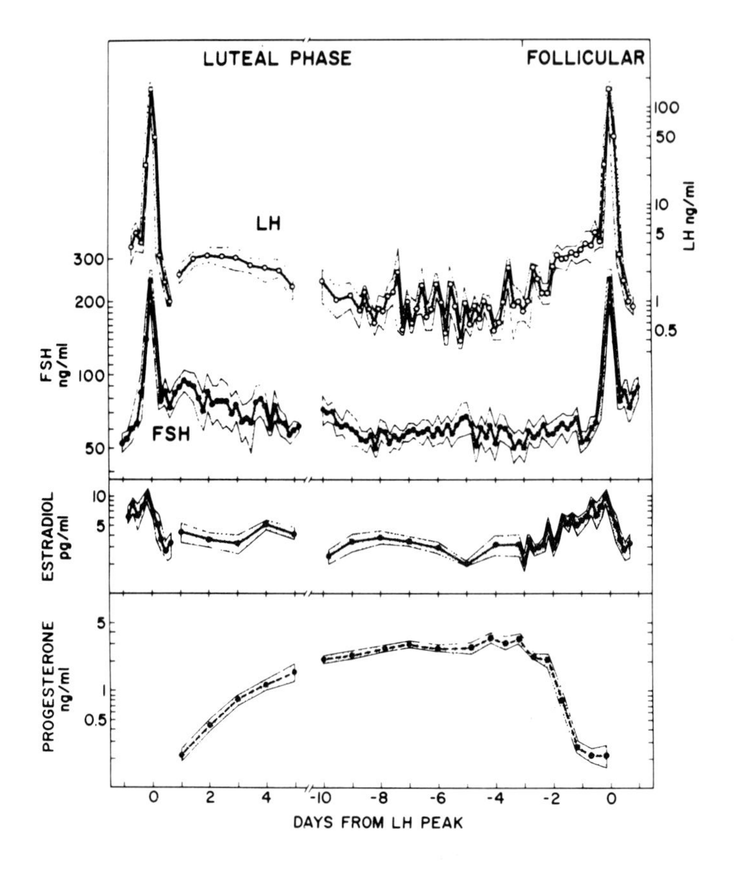

Fig. 11 *Serum concentrations (mean + SEM) of FSH, LH, oestradiol and progesterone during the oestrous cycle.* (Note the logarithmic scale.) [From Goodman *et al.* (1981) by permission of The Williams and Wilkins Co.]

oestrogen of near the size of the oestrus peak also occur during the luteal phase due to development and then astresia of follicles.[309] The progesterone level, negligible at oestrus, rises to about 1 ng/ml at about day 4. Peak levels during the oestrous cycle and early pregnancy (about 4 ng/ml) are low compared with those in many other species. Experiments involving the insertion of intra-uterine devices on different days of the cycle indicate that the systemic progesterone level will have reached a sufficient threshold to have a physiological effect on the uterus on day 4 of the cycle[404] (see below).

Table 21 Concentrations of some reproductive hormones.

Hormone	Concentration	Comments
Anterior pituitary gland		
FSH	62 ng/ml	Ewes, luteal phase[286]
	171 ng/ml	Ewes, at LH peak[286]
	133 ng/ml	Ewes, 24 h after LH peak[286]
LH	2·6 ng/ml	Ewes, mid luteal phase[286]
	75 ng/ml	Ewes, 9 h after onset of oestrus[286]
	2·5 ng/ml	Rams, breeding season[321]
	1·8 ng/ml	Rams, non-breeding season[321]
Prolactin	41 ng/ml	Rams[104]
	13 ng/ml	Wethers[104]
	50 ng/ml	Ovariectomized ewes[102]
	292 ng/ml	Lactating ewes[102]
	37 ng/ml	Pregnant ewes[102]
	27 ng/ml	Anoestrous ewes[102]
Ovary		
Oestradiol 17β	11 pg/ml	Ewes, luteal phase[286]
	21 pg/ml	Ewes, 8 h before oestrus[286]
Progesterone	3·7 ng/ml	Ewes, mid-luteal phase[286]
	2 ng/ml	Ewes, mid-luteal phase[94]
	3 ng/ml	Ewes, 70 days pregnant[94]
	12 ng/ml	Ewes, 120 days pregnant[94]
Testes		
Oestradiol 17β	5·4 pg/ml	Rams, breeding and non-breeding season[321]
Testosterone	5·2 ng/ml	Rams, breeding season[321]
	1·2 ng/ml	Rams, non-breeding season[321]

Some progesterone is converted into 20 dihydroxyprogesterone. The levels of this hormone are similar to and follow those of progesterone.[30]

At 2 days before the next oestrus, the progesterone level drops precipitously due to cessation of secretion by the corpus luteum and this cessation is caused by the release of prostaglandin $F_{2\alpha}$ from the uterus. The prostaglandin is released from the uterus into the uterine vein and some of it passes by diffusion across the venous wall, through the wall of the closely adherent ovarian artery and thence into the ovarian arterial blood. The mechanism is not entirely clear and some consider that prostaglandin E_2 released from the uterus of pregnant ewes at the same time as prostaglandin $F_{2\alpha}$ may have an antiluteolytic effect.[122]

There is evidence for a "clock" mechanism that controls the length of the luteal phase of the oestrous cycle. This evidence is based principally on the cycle shortening effects of exogenous progesterone given early in the cycle as there is a relatively constant interval of 12 ($\pm$1 S.D.) days between the first dose of progesterone and return to oestrus. It is believed that the rise in progesterone in the normal cycle to a physiologically effective level on about day 4 starts the clock.[53] Continuous infusions of progesterone had no effect on cycle length when given early in the cycle into the carotid or ovarian arteries but shortened the cycle when given in minute amounts into the uterine arteries, indicating that this clock is uterine.[53]

The LH peak occurs about 26 h after the onset of oestrus. Most eggs are released at 12–24 h after the onset of oestrus, or 23–25 h after the LH peak.[97] The median interval between twin ovulations is 1·2 h.[396] Eggs remain in a fertilizable state for at least 12–18 h.[222] Most ova have passed into the uterus by 66 h after ovulation by which time they are in the 8-cell stage.[191] It is not clear if capacitation occurs in sheep—if it does, it must occur rapidly.

Maternal recognition of pregnancy occurs at about day 12 of the cycle. This day is critical as, if the embryo/s are removed before it, the ewe will have a cycle of normal length whereas if removed after it, the cycle length will be extended by several days due to failure of the corpus luteum to regress at the normal time.[273] Also with transfer of an embryo from a donor ewe to a recipient ewe which was in synchrony with the donor ewe, the embryo is likely to develop if the transfer is done before day 12 but not if done after this day.[273] It is thought that the signal from the embryo to the ewe is a substance, possibly a protein[263] or an oestrogen. It may be significant that at about day 12 the embryo elongates from a sphere of about 1 ?m to a cylinder of about 10 cm in 24 h.

Implantation is a gradual process occurring between about days 15–30. The placenta is cotyledonary within the order of 100 cotyledons. In single pregnancies, the functional cotyledons are in approximately equal numbers in both horns although the lamb will be in one horn. Histologically, the six layers of the syndesmochorial type of placenta are present.

Pregnancy lasts for slightly longer in Merino than in British breed sheep, being about 150–154 days in the former and 144–148 in the latter. After about 40 days, the blood progesterone level commences a steady rise due to the placental production of this hormone and it reaches a peak shortly before parturition. After 40 days, the ovaries can be removed without inducing abortion as adequate progesterone is produced by the placenta by this time. Oestrogen levels are low until the end of pregnancy with the main oestrogen being oestrone sulphate. Oestriol is virtually undetectable. Ovine placental lactogen (chorionic somatomammotrophin) is first detectable at about 60 days and rises to a maximum at about 100 days before falling slightly until parturition.[168] Further details of hormone changes during pregnancy are given by Cox.[94]

Testosterone administered to pregnant ewes as implants will masculinize the genital tubercule of ewe lambs if given at around 40–50 days.[401] If given slightly later, ewe lambs will adopt the male urination position.[81] Normal cycling after expected puberty will be inhibited in many of these ewe lambs, especially if the testosterone is given to their mothers early.[82] The critical time for affecting hypothalamic differentiation appears to be between days 50 and 80 of gestation.[401]

The processes involved in the initiation of parturition have been well investigated,[236] especially as it was believed that an understanding of the processes involved in the ewe might help to elucidate processes in women. The scatter in gestation lengths of ewes is low (S.D. about 2 days) except in ewes in the western United States which have grazed the weed *Veratrum califorincum*. Lambs in these ewes had prolonged gestation lengths and usually eventually died *in utero*. Investigations showed the lack of development of the head region and in particular of the anterior pituitary gland.[127] Also the adrenal cortex was poorly developed. By following these clues, a considerable amount of research showed the following sequence of events. The timing of the day of birth is determined by the foetal brain which at about one week before parturition increases the rate of secretion of ACTH from the foetal pituitary gland. Until this time, this secretion rate has been minimal. The ACTH causes hypertrophy of the adrenal cortex and markedly increases the level of cortisol in the foetal blood. Cortisol stimulates production of prostaglandin $F_{2\alpha}$ from foetal membranes. The cortisol and/or the prostaglandin stimulate secretion of oestradiol 17β and reduce secretion of progesterone from the corpus luteum and placenta. The prostaglandin, aided by the increased oestrogen and the removal of the "progesterone block", increases uterine motility and probably also causes cervical dilation (see p. 6). These events are most evident in the 24 h prior to birth. It is possible that during parturition, oxytocin augments the release of prostaglandin $F_{2\alpha}$.[138] This knowledge has led to methods for induction of parturition. ACTH does not cross the placenta but synthetic adrenal corticosteroids do cross readily.[28] Betamethasone (16 mg/kg, i.m.[294]) will induce parturition in almost all ewes in about 48 h from a gestational age of 130 days onwards.

Much research has been done on culture of early ovine embryos and on ovum and embryo transfers.[311] By stimulation of ovum development with a suitable dose of pregnant mares serum (PMS, about 1200 i.u.), about 12 fertile embryos can be obtained. Success rates with embryo transfers can be over 80%.[399] Success has also been achieved with freezing of ovine embryos.[400] A recently developed technique shows considerable promise as an aid to several areas of experimentation as it has been possible at the two cell stage after fertilization to split the zona pellucida, remove and separate the blastomeres and then seal them back into separate zona pellucidas. These can then be put back into one or two ewes so as to produce identical twins at will (if put into two ewes, the strange situation arises of identical twins with different mothers). In the first trial with this technique, five sets of twins were produced from 16 ewes.[398]

Intra-uterine devices have been used in sheep for research purposes. The type usually used is a spiral polyethylene coil about 5 cm long and 1 cm in diameter which when inserted into a uterine horn causes distension. The distension results in continuous production of prostaglandin $F_{2\alpha}$ which prevents a corpus luteum from becoming functional. Cycle length is considerably reduced as there is no lueteal phase but only a series of follicular phases as in the rat. If a smaller intra-uterine device such as a thread which does not distend the horn is inserted, normal cycles occur.[178] These threads which are still contraceptive are equivalent to the type of device used in smaller laboratory animals. If the distensible type of device is filled with slow-release progesterone, the local presence of progesterone overcomes the luteolytic effects of the distension.[278]

Lactation in sheep shows no unusual features. Composition of sheep milk is shown in Table 22. Artificial induction of lactation can be produced by priming ewes with oestrogen for a month, giving a prostaglandin analogue and then hand milking.[143]

Table 22 Composition of milk.

Constituent	Milk	Colostrum
Dry matter (%)	15–19·3[50]	21·1[50]
Fat (%)	6·5–7·5,[293] 9·1–10·9[295]	12·4–13·0[295]
Solids not fat (%)	11·2[293]	
Protein (%)	5·4,[293] 4·3–5·2,[50] 6·5–7·0[295]	8·0,[50] 10·8–11·8[295]
Casein (%)	3·27–4·03[50]	4·73[50]
α lactalbumin (%)	0·18–0·26[50]	0·45[50]
β lactalbumin (%)	0·25–0·35[50]	0·8[50]
Proteose, peptones (%)	0·04–0·16[50]	0·39[50]
Globulins (%)	0·32–0·39[50]	1·06[50]
Non-protein nitrogen (%)	0·08–0·21[50]	0·54[50]
Lactose (%)	5·0,[293] 4·0–4·1[295]	3·3–3·4[295]
Ash (%)	0·8,[293] 0·88–0·94[295]	0·86–0·95[295]
Energy (mCal/kg)	1·1[293]	

Values from milk from [50] and [295] are the range over weeks 1 to 12 and 1 to 16 respectively.

The vascular cone on top of the testis of the ram is impressive as a 200–400 cm length of testicular artery is compressed into a 8–14 cm length and surrounded by venous plexuses. The cone has two functions. It substantially reduces the arterial pulse pressure (probably mainly in the top third) and also acts as a heat exchange mechanism (probably mainly in the lower two thirds)[383] with the result that the testicular temperature is maintained at about 34·5°C.[141]

This counter current mechanism also acts to increase the intra-testicular concentration of steroids such as testosterone and so dampen fluctuations in levels in the seminiferous tubules.[378] A similar counter current mechanism increases the concentration of progesterone in an ovary with a corpus luteum[384] and probably also of other ovarian steroids.

3.9 Digestive System

The ruminant digestive system is probably the system that provides the greatest contrast to the typical laboratory animal. The sheep is a ruminant and the following discussion is an outline of the more important features of ruminant digestion with cited values from experiments using sheep. For more information, the interested reader should consult Church,[79] Hungate[199] and the series of International Ruminant Symposia which have been held every four or five years.[116,234,246,297,313]

A. Salivation

Saliva is secreted from several sets of glands of which the main ones are the parotid, submaxillary, sublingual and inferior molar glands (see Table 23). The submaxillary glands secrete only during feeding whereas the others secrete at a slow continuous rate which increases during feeding. Most attention has concentrated on the parotid glands due to their size, volume of secretion and ease of sampling that secretion. Saliva is devoid of digestive activity. The parotid saliva in particular is a very important source of fluid to the rumen and the large amounts of bicarbonate and phosphate in the parotid, palatine, buccal and pharyngeal glands are important in buffering the acids produced by fermentation in the rumen. Stimulation of the oesophagus and parts of the reticulo-rumen will markedly increase saliva flow. Diversion of saliva from one of the parotid glands to the exterior results in a rapid depletion of sodium and a depression of appetite unless the animal is given access to a salt lick or saline solution to drink.[1] The sodium to potassium ratio in parotid, submaxillary and inferior molar glands changes from high sodium: low potassium to low sodium: high potassium but there is no change in the composition of the other glands. The changes in mixed saliva are reflected by changes in rumen cations.

Table 23 Concentrations of electrolytes and pH in some secretions, cerebrospinal fluid and plasma.

	Parotid saliva[219]	Submaxillary saliva[219]	Bile[67]	Pancreatic juice[350]	Cerebrospinal fluid		Plasma[29]
	Mean (range)	Mean (range)	Mean	Range	Mean[29]	Mean[276]	Mean
Sodium	170 (147–185)	9(3–16)	150	135–165	150	150	163
Potassium	13 (6–31)	16 (10–25)	4·4	3·9–5·4	3·2	2·8	4·6
Calcium	—	—	—	4·0–5·7	1·2	1·2	2·5
Magnesium	—	—	—	0·7–1·5	0·9	0·9	0·8
Chloride	11 (9–16)	11 (7–15)	118	110–126	123	131	112
Bicarbonate	112 (103–125)	9 (5–14)	23	15–30	—	24	—
Phosphate	48 (25–64)	5 (2–10)	—	—	0·4	—	2·07
pH	—	—	—	7·2–7·8	7·39	7·45	7·42

Units (except pH) mEq/1.

B. Oesophagus

The oesophagus is unusual in having not smooth muscle but striated muscle which is innervated by the vagus nerve. Peristaltic waves associated with swallowing pass at a rate of about 20–30 cm/s while those passing in the reverse direction (associated with regurgitation and eructation) travel more rapidly (170–200 cm/s).[117]

C. Rumen

The rumen (together with the reticulum) is a large microbial fermentation chamber in which the food is mixed, kept moist and maintained at about 40°C in a highly anaerobic buffered environment. The microflora is highly specialized and consists of numerous species of bacteria and protozoa. Most are unique to the rumen. One species, previously believed to be a protozoal species was only recently recognized as an anaerobic fungus.[25] Bacteriophages are also present.[193] Numbers vary considerably depending on factors such as the diet and frequency of feeding. The protozoa (10^5–10^6/ml of fluid) are much less numerous than the bacteria (10^9–10^{10}/ml) but being larger can represent as large a volume of microbial cell mass as the bacteria. Depending on species, each protozoan can ingest in the order of 100 bacteria/min.[88]

Considerable research has been done on the functions of individual species of micro-organisms[199] but a common approach has been to consider the rumen contents as a "tissue" with an unusual digestive function. In brief, carbohydrates including cellulose, hemicellulose, pectins and starches are fermented after hydrolysis to form volatile fatty acids of which acetic acid predominates but propionic and butyric acids are also found in significant quantities. Soluble proteins are hydrolysed to amino acids but free amino acids are found only in trace amounts as they are either taken up into organisms or are deaminated to ammonia and short chain organic acids. This is the main source of the small amounts of valeric, isobutyric and isovaleric acids found in rumen contents. All of these are volatile fatty acids as they can be steam distilled from digesta. For synthesis of protein, the rumen microbes use some of the free amino acids and also ammonia and short chain acids. Insoluble proteins are not degraded in the rumen as the microbial proteases can act only on soluble substrates.

Fats are hydrolysed to glycerol (which forms propionic acid) and free fatty acids which are hydrogenated and bound to food particles.[170] The predominant fatty acid in plants is linolenic acid (60–70% of plant lipid[106]) which tends to be found as monogalatosyl- and diglactosyl-diglycerides.[244] Hydrogenation can take several pathways and a variety of products can be found, but the *cis*, *cis*, *cis* isomer of linolenic acid tends to be converted to an 11 *trans*, 15 *cis* diene isomer and then to an 11 *trans* isomer as successive double bonds are removed. Hydrogenation of *trans* mono-enoic acids is slower than of *cis* isomers and as a result some of the *trans* monosaturated fatty acids pass to the duodenum, are absorbed and are stored in depot fat.

Significant amounts of trimethylamine can be found in the rumen. Some of this is formed from methyl groups of choline in dietary phosphatidylcholine compounds.[280] Trimethylamine is degraded to methane but after a meal its rate of degradation is slower than its rate of formation and hence it can accumulate for a time.

Nucleic acids which on a dry weight basis are 5–10% of the nitrogenous compounds in grass are hydrolysed to purine and pyrimidine bases. These may be absorbed from the rumen as they disappear from the rumen faster than a marker.[342]

Many of the vitamins used by sheep are synthetized in the rumen by the microflora. Vitamins A, D and E are not synthetized and sheep depend for these on dietary sources. Comparatively little dietary vitamin E is hydrogenated in the rumen.[106]

There is a relation between the amount of microbial protein synthetized and the quantity of organic matter fermented in the rumen. Experiments have given values for this relation of 8–25 g per 100 g of organic matter.[342]

Rumen fermentation produces large volumes of gas which is principally carbon dioxide but which also contains methane and a little nitrogen. The gas is regularly eructated to avoid overdistension of the rumen although there is always a gas bubble on top of the digesta. For some unknown reason, most of the eructated gas passes from the oesophagus to the trachea and enters the lungs before being exhaled.[118] The condition of bloat caused by the formation of a stable foam which cannot be eructated is common in cattle but rare in sheep. The methane formation can be up to 8% of the feed energy and represents a significant loss. It can be reduced by adding certain additives including unsaturated fatty acids to the feed.[108]

Proper functioning of the rumen depends on regular mixing of the contents and this is done by complex series of movements which occur once or twice per minute and which are initiated and coordinated by the vagus nerve. There are two basic types of movements which have been termed *A* and *B* sequences. In the *A* sequence which is the more common, there is a double contraction of the reticulum during the latter part of which a wave of contraction passes over the rumen starting in the cranial blind sac and passes caudally and ventrally to finish in the posterior ventral blind sac. The *B* sequence passes in the reverse direction and does not involve the reticulum. Modifications of these waves of contraction move digesta as necessary towards the cardia for regurgitation and the gas bubble at the top of the rumen towards the cardia for rumination. At each contraction, a little fluid and finely divided digesta pass from the reticulum into the omasum.

The rumen has a highly keratinized epithelium which has a high blood flow and reasonable permeability so as to allow considerable absorption from the rumen contents. Many papillae increase the absorptive surface. Volatile fatty acids and ammonia are absorbed by diffusion in the un-ionized state. Hence the absorption of these substances is pH dependent with the rates of absorption being greater in the acid rumen environments found during rapid fermentation of food. Metabolism of much of the butyric acid to ketone bodies (mainly $D(-)\beta$-hydroxybutyrate but also some acetoacetate) and propionic acid to lactate occurs in the rumen wall but little acetate is metabolized. Hence the portal blood contains a higher proportion of acetate than does rumen fluid. Much of the sodium which enters in saliva is absorbed by an active transport mechanism and water follows passively. However, soon after a meal, the products of fermentation and the release of potassium from crushed plant cells causes the osmotic pressure to rise in the rumen and water can pass from blood into the rumen (see p. 85).

D. Omasum

Due to the difficulties in obtaining surgical access to the omasum, less is known about the function of this organ. Some of the digesta from the reticulum passes directly through to the abomasum while the remainder is held between the omasal leaves for absorption of water, ammonia and volatile fatty acids.[127]

E. Abomasum

The digesta which enters the abomasum or true stomach is different from that which enters the stomach of monogastric animals in that it has had most of the readily available food nutrients removed by fermentation, is in a finely divided ground state, contains large amounts of microbial cell material and, in the grazing animal at least, enters at a relatively continuous rate. In laboratory sheep fed only once each day as many laboratory sheep are fed, the even flow rate will continue for some 8–12 h after which time it decreases. The protein content of the digesta will be reasonably constant in composition as much of it will be the products of the rumen microflora (Table 24). In addition to the microbial protein, there will be any undigested (insoluble) feed protein, microbial nucleic acids and microbial cell wall nitrogenous compounds. The microbial nucleic acids constitute 12–18% of the total microbial nitrogen and 75–90% of that synthetized in the rumen is absorbed in the intestines. However 25% of that absorbed is excreted in the urine as allantoin and the remainder as urea and uric

Table 24 Amino acid composition of rumen bacteria and protozoa.

Amino acid	Bacteria[192] g/100 g cells	Bacteria[301] g/100 g protein	Protozoa[301] g/100 g protein
Alanine	2·9	5·9–6·5	3·9–5·2
Arginine	1·9	5·1–6·1	4·3–7·6
Aspartic acid	4·6	11·1–11·8	12·3–16·1
Diaminopimelic acid	Trace	0·8	
Glutamic acid	4·8	11·9–14·2	12·5–18·0
Glycine	1·9	5·1–6·1	4·1–5·0
Histidine	0·7	2·0–6·3	1·8–4·3
Leucine	3·2	4·7–7·3	7·3–8·1
Isoleucine	2·2	3·6–6·4	7·3–8·3
Lysine	2·9	7·8–9·7	9·2–13·1
Methionine	1·3	2·6–3·0	1·4–2·2
Phenylalanine	2·4	4·9–5·1	3·8–6·2
Proline	1·4	3·6–5·3	2·1–4·5
Serine	1·8	3·8–4·9	3·6–6·2
Threonine	2·3	4·9–5·4	4·7–5·8
Tyrosine	2·0	4·2–4·7	4·1–5·4
Valine	2·3	4·2–6·7	4·0–5·4

acid.[8] Therefore the nucleic acids are highly digestible but of questionable value to the animal. Cell wall nitrogen is in the form of mixtures of *D*- and *L*-amino acids, unusual amino acids such as diaminopimelic acid and amino sugars, and some of these substances are also of questionable value to the animal. Thirty per cent of microbial cell walls is in the form of amino acids and this fraction is 15% of the total cell mass. Estimates from rat-feeding experiments place the biological value of rumen microbial nitrogen at 54–70% (most estimates 62–69%[301]).

Little sugar or starch-like storage polysaccharide reaches the abomasum unless large amounts of starch are fed when some escapes rumen fermentation by leaving the rumen soon after feeding. Of the little that does normally reach the abomasum, much is as a storage polysaccharide in protozoa.

Abomasal function is similar to gastric function in non-ruminants. The cephalic phase of secretion is unimportant and the main stimulus to secretion is the amount of volatile fatty acid in the contents entering the abomasum. The volume of secretion of the fundic gland region is 2–8 litres/day and of the pyloric region 0·08–0·12 litres/day. The pH of contents is typically 2·5 and secretion is inhibited if the pH is lowered to pH 2·0.[12] Pepsin is present. An important role of the hydrochloric acid is to kill the micro-organisms and rupture the cell walls so as to release cell contents.

F. Liver and pancreas

Bile is secreted at a rate of 0·5–1·5 litres/day and pancreatic juice at 0·3–0·4 litres/day.[68] A feature of these secretions is the comparatively low secretion of bicarbonate (e.g. pancreatic bicarbonate concentration in dogs is about four times that in sheep[350]). Unlike most other animals, bile is a more important source of buffering for abomasal acid than is pancreatic juice. The gall bladder is contractile and will concentrate bile in the fasted animal but it does not play an important role in an animal which is fed frequently as most bile passes directly to the duodenum.[68]

The normal gut hormones, secretin and cholecystokinin-pancreozymin appear to regulate secretion from these two organs.[68]

Table 25 Hepatic parameters.

Parameter	Value
Clearance Sulfobromophthalein (BSP)	54 ml/kg/min[216]
Halflife Sulfobromophthalein (BSP)	3·2 min[93]
Halflife Indocyanin green	4·8 min[93]
Halflife Rose Bengal	5·6 min[93]
Halflife Bilirubin	8·7 min[93]

G. Small intestine

The abomasum contracts every 9–12 s, slowly filling the duodenal bulb. The initial passage of digesta through the duodenum is very rapid, occurring as a "rush" to the transverse jejunum.[394] Passage through the remainder of the small intestine takes 2·25–4·5 h.[91] Passage through the large intestine is slower (10–26 h) and is dependent largely on the level of food intake.[91,159] The migrating myoelectric complex in the small intestine consists of a phase of irregular spiking activity followed by a phase of regular spiking activity and then a quiescent period.[312] Segmental contractions are signalled by regular bursts of spike potential lasting a few minutes. Most, but not all waves of segmental contraction reach the ileum, taking 1·5–2 h to do so.[158] Electromyography has shown that post-operative depression of intestinal motility (ileus) is more severe in sheep than in dogs and less marked in the duodenum than in lower parts of the tract.[63]

Small intestinal digestion is basically similar to that in monogastric animals. One difference is that the contents are not completely neutralized in the duodenum and reach pH 7 only in the ileum. Hence most of the digestion and absorption is done at a slightly acid pH. Of the nitrogenous material entering the abomasum, some 60–70% is digested in the small intestine and 10–20% disappears in the large intestine largely as ammonia.[175]

Intestinal fat digestion probably shows the most interesting differences from monogastric digestion.[231] The fat is already hydrolysed and so the need for pancreatic lipase can be questioned. However fat absorption is depressed if pancreatic juice is diverted from the intestine. It is also depressed if pancreatic juice is prevented from mixing with bile in the common bile duct before entering the duodenum. In the common bile duct, it appears that the phospholipase in pancreatic juice converts bile lecithin into lysolecithin which is a better detergent than lecithin. The lysolecithin therefore induces the formation of micelles from the fatty acids bound to food particles and so assists in absorption of the fat.

Maltase is the most abundant carbohydrase in the small intestine but amylase and lactase and smaller amounts of sucrase and cellobiase are also present. Amylase levels are possibly much lower than in the dog.[238] Activities tend to be highest in the jejunum.[181] Trypsin, chymotrypsin and carboxypeptidase A are most active in the second quarter of the small intestine[151] as are dipepidases.[347] The low pH probably has a limiting effect in the first quarter. Large amounts of nucleases are secreted into the intestine in pancreatic juice.[342]

Absorption of sodium and of water is quantitatively greatest in the region 0 to 7 m (small intestinal length is about 25 m) whereas absorption of nitrogenous compounds occurs in the region 7 to 15 m.[34] Fat absorption is also in the upper jejunum and it is noteworthy that this occurs in an acid environment which inhibits fat absorption in humans.[231] Microbial action becomes significant in the lower part of the small intestine.[34]

H. Large intestine

Similar types of processes to the rumen occur in the large intestine[374] but normally less than 10% of digestible food material reaches the large intestine. Under some circumstances such as feeding finely ground feeds, this can rise to 25% due to the rapid escape of this material from the rumen before fermentation is complete. Then digestion in the large intestine becomes quantitatively important.[84] The other important function of the large intestine is the absorption of water and electrolytes allowing the excreta to be passed in the form of small hard pellets containing about 40% dry matter.[176] As a consequence of the large amount of indigestible dietary plant material, the volume of faeces is large and is responsible for obligatory losses of nitrogen (partly as microbial cell walls) and other substances (see Table 27). Hence digestibilities of food constituents are lower than might otherwise be expected.

3.10 Metabolism

For a great range of sizes of animals (mice to elephants), the fasting metabolic rate is approximately 70 kcal/unit of metabolic size (weight raised to the power of 0·73 ($kg^{0·73}$)).[44] Mature sheep appear to have a lower fasting metabolic rate than this relation suggests. Estimates indicate that the fasting metabolic rate is 55–59 kcal/$kg^{0·73}$ which is about 15% lower than expected. It might be assumed that the large bulk of digesta in the rumen and intestines present even after 24 h of fasting would add non-metabolic weight to the animal and so cause a low value but estimates for cattle which also have a large bulk of digesta are about 15% higher than the above formula suggests.[44]

Table 26 Metabolic parameters.

Parameter	Value
Maintenance energy	1800–2120 kcal/day[134]
Maintenance energy	47–62 kcal/kg/day[134]
Fasting metabolism	70·5 kcal/kg 0·73 weight[44]
Requirement for horizontal locomotion	0·69,[134] 0·59[44] cal/m/kg
Requirement for vertical locomotion	7·64,[134] 6·54[44] cal/m/kg
Requirement for feeding	0·3–1·9 cal/kg/g ration[411]
Requirement for rumination	0·34 kJ/kg/h[356]
Requirement for standing up	47 kJ/kg[356]
Requirement for standing (compared with lying awake)	0·7 kJ/kg/h[356]
Respiratory quotient	0·95 (fed),[210] 0·92 (fasted),[210] 0·64–0·69[165]

Maintenance energy expenditure and the energy costs of various types of "work" are shown in Table 26. The maintenance metabolic rate is in the order of 50% greater than the fasting metabolic rate. Changing the environmental temperature to one above or below the thermoneutral zone will increase the metabolic rate (Fig. 12).

In a general comparison with a monogastric animal such as man, the sheep eats more and excretes more than man (Table 27). Its food contains a considerable amount of energy but is much less digestible than a typical human diet. The rumen microbes help the sheep to digest plant fibre but in doing so they use extra energy. These comments should be kept in mind when interpreting information given on p.79.

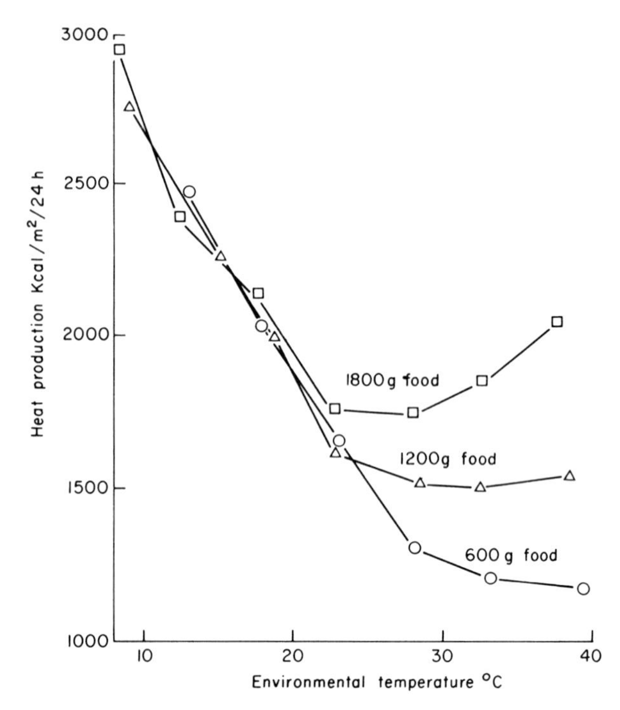

Fig. 12 *Heat production at different environmental temperatures of closely clipped sheep given different amounts of food.*
[From Blaxter by permission of Hutchinson Co.]

Table 27 Comparison of the energy balance in a grazing sheep with that of a typically over-fed man.[80]

	Sheep (mJ/day)	Man (mJ/day)
Amount of energy taken in	41	16
Amount of energy lost		
in excreta	21	1
as heat, resting	5	5
eating	2	1
walking 8 km	2	2
metabolism of food	6	2
Energy stored	5	5

A. Glucose

Unlike monogastric animals, the main sources of energy are volatile fatty acids. Glucose is essential for proper functioning of the tricarboxylic acid cycle and for the synthesis of a variety of compounds. In most tissues, glucose is not used for energy but it is the substrate for metabolism in the central nervous system, the pregnant uterus and the lactating mammary gland.[24] In consequence, glucose makes only a minor (4–11%) contribution to respiratory carbon dioxide and possibly only 30–35% of the glucose that is produced is oxidized.[238]

The glucose entry rate is in the order of 109 g/day on a typical ration.[56] In a fasted sheep, the production drops to about 140 mg/kg$^{0.75}$/hour.[238] Compared with monogastric animals, comparatively little glucose is recycled through metabolic intermediates.[56] Also in contrast to monogastric animals, the rate of gluconeogenesis decreases with fasting.[238]

While the brain normally metabolizes glucose, it is disputed as to whether it can metabolize ketone bodies as can the brain of man.[210,240] The testis is another organ which uses mainly glucose.[329]

Much (54%) of the glucose is formed from propionic acid (or lactic acid produced by metabolism of propionic acid in the rumen wall). Some 11–17% is produced from amino acids of which glutamate and alanine are the principle precursors. Valeric and isobutyric acids, the small amount of glycerol which is absorbed and the pentoses from nucleic acids are other sources.[238]

The plasma glucose concentration is lower than in monogastric animals. The high levels in lambs fall to adult levels by 3 months of age. It has been suggested that this fall may be associated with the development of the adult rumen[230] but various experiments have shown that it is independent of rumen development or of the increased levels of volatile fatty acids (see [230]). Unlike monogastric animals and the foetus, adult red blood cells contain negligible glucose and hence sheep blood glucose levels are misleading.[194]

The glycolytic activity of sheep blood is lower than for other species. The white cells have some activity while the activity of red cells is negligible.[239]

Glucose tolerance tests indicate that the young lamb has a response that is similar to that of the normal human whereas the response of adult sheep is slow and like that of diabetic humans. This is not unexpected in a species that rarely encounters high blood levels of glucose. Young lambs have a glucose response to injected insulin similar to that of monogastric animals whereas in adult sheep the fall in plasma glucose is much slower, probably because the liver and fat cells cannot use glucose for lipogenesis (see below). The blood glucose level of adult sheep must drop below 5 mg/100 ml before coma ensues.[304] These comments suggest a lack of importance of insulin in the adult sheep. However diabetes produced by pancreatectomy or administration of alloxan causes the typical severe symptoms seen in monogastric animals.[24,204]

Unlike monogastric animals, there is little relation between the plasma glucose level and the concentration of insulin or the insulin secretion rate.[23] The reason is that all of the volatile fatty acids produced by fermentation in the rumen except acetate are up to 15 times more powerful on a molar basis as stimulators of insulin secretion than is glucose.[197] Paradoxically, these fatty acids also cause hyperglycaemia,[296] probably through stimulation of glucagon secretion. Their precise role in the control of these two hormones is unclear as the liver removes most of the fatty acids absorbed from the gut with the exception of acetate and hence blood levels of the higher volatile fatty acids are not closely related to the absorption rate. (Also paradoxically, acetate utilization depends on insulin.[23])

As well as direct stimulation of insulin secretion by substances perfusing the pancreas, there is evidence for nervous control as both feeding and sham feeding result in a doubling in insulin level within 5 min of feed being offered.[22]

There is little uptake of glucose from blood by the liver. The activity of glucokinase in sheep liver is very low, as are the activities of many other enzymes involved in carbohydrate metabolism such as glucose-6-phosphate dehydrogenase, fructose-1,6-diphosphate aldolase and glycerol-3-phosphate dehydrogenase. However the liver is well equipped with the enzymes necessary for the release of glucose from stored glycogen and for the synthesis of glucose from propionate.[230] It may be that glucokinase is the limiting enzyme for utilization of glucose as the liver is able to take up large amounts of infused traces of *D*-galactose and smaller amounts of *D*-fructose[194] which require different hexokinases.

The liver is more sensitive to glucagon than to adrenaline.[23]

B. Lipids[26,231]

Because of the hydrogenation of unsaturated fats by the microflora in the rumen, the sheep has only a meagre supply of polyunsaturated fatty acids. The depot fats are mainly saturated but may contain a variety of odd- and branched-chain fatty acids formed as a result of microbial action in the rumen.[230] A diet with high levels of barley grain promotes the accumulation of branched-chain fatty acids[337] because this type of diet produces large amounts of propionic acid of which some escapes metabolism in the liver and is incorporated into fatty acids in adipose tissue causing branched chains to form. High levels of such branched chain fatty acids can result in the fats of the carcase not setting at room temperatures.

The polyunsaturated fatty acids that are present in sheep are mainly as esterified lipids in plasma. In spite of the small amounts of linoleic and linolenic acids absorbed,[230] there is little evidence for essential fatty acid deficiency. The foetus is in a particularly perilous state for essential fatty acids as the placenta is more permeable to free fatty acids which are mainly saturated than to esterified fats but again there is little evidence for fatty acid deficiency in the foetus. A placental desaturase enzyme may assist the foetus here.[331]

In man, the liver is the main site for gluconeogenesis and of lipogenesis but the liver in the sheep plays little role in lipogenesis which takes place mainly in adipose tissue and the mammary gland of the lactating ewe. Glucose is limited in supply and so virtually no glucose is used for fatty acid synthesis. Acetate is mainly used instead. β-hydroxybutyrate can also be used for the first four carbon atoms but not for other parts of the fatty acid molecule. The biochemical reason why glucose is not used is the lack of the enzyme ATP-citrate lyase in the cytosol which in non-ruminants allows citrate formed from glucose via the tricarboxylic acid cycle to be converted to acetyl Co A and incorporated into fat.

Fatty acid synthesis requires the generation of NADPH. This reducing compound is generated from both the pentose phosphate cycle and the isocitrate cycle but not from the malate trans-hydrogenation cycle due to negligible levels of NADP-malate dehydrogenase in the cytosol. The isocitrate cycle uses acetate metabolism and the triose fragments formed in the pentose phosphate cycle are extensively recycled. Hence the NADPH is generated with the use of minimal amounts of glucose.

In addition to fat synthetized in the body, depot fats contain proportions of *trans* monounsaturated fats (p. 71) and of phytic acid which originates from the chlorophyll from feed. The internal fat depots (perinephric and mesenteric) are far richer in *trans* isomers than subcutaneous depots.[106]

Regulation of lipogenesis appears to be by alteration of the levels of acetyl Co A carboxylase, the rate limiting enzyme. Lipolysis in monogastric animals is induced by adrenaline and/or glucagon by regulating the amounts of adenyl cyclase in cells. Both of these hormones induce lipolysis in ruminants but neither has a marked effect.

C. *Amino acids*[10]

The main source of protein to the sheep, that synthetized in the rumen by the microbes, is relatively constant in composition and is reasonably well balanced for the sheep's neeeds (Table 24). Surplus amino acids are deaminated in the liver and most of the ammonia is converted to urea. The blood urea concentration which reflects the rate of deamination of amino acids provides an indication of the protein status of a sheep. As wool contains large amounts of sulphur-containing amino acids, these are most likely to be limiting.[281]

As in monogastric animals, glutamine is used as an ammonia carrier to the kidneys, but because the kidneys have a net release of glutamine during feeding, there are other carrier systems of which one is the arginine-citrulline system.[37] This is a reason for the high level of citrulline in blood (Table 13).

Kinetic studies show that there is always net hepatic removal of amino acids which are normally supplied by intestinal absorption but which in the fasted animal come from peripheral tissues including muscle.[180]

D. *Steroids*

Metabolism of steroids shows some differences from man. Although traces of numerous metabolites can be found, reproductive steroids tend to be converted into 17α-(oestradiol,[362] testosterone and androstenedione[409]) or 20α-(progesterone[361]) hydroxysteroids. With some steroids (progesterone[361]) there is a 5α-reduction. Small amounts of 20β-progesterone and pregnane-3α,20α-diol are also found in blood of pregnant sheep.[94] Androgens can be mono- and/or di-conjugated with sulphate and/or glucuronic acid although 80–90% are monoconjugated, and are excreted in either bile (allowing metabolites to be found in faeces) or urine.[409] Progesterone[361] metabolites are also found in urine but oestradiol metabolites are mainly in faeces following biliary excretion.[407]

Oestriol is conjugated at the 3 position with either sulphate or glucuronic acid with the glucosiduronate being mainly excreted in bile and the sulphate in urine. There is evidence for an enterohepatic circulation of oestriol with the sulphate conjugate being hydrolysed in the small intestine and the 3 glucosiduronate reabsorbed unchanged.[271] There is no double conjugation of oestriol as in humans.

Oestrone sulphate is the most abundant oestrogen in pregnant sheep but there is a rapid reduction to oestrone-17α.[362] In the pregnant sheep, oestriol is not detectable and unconjugated oestrogens are detectable only in the last 2–3 days.[360]

Sites of location of steroid biosynthetic enzymes in the uterus and foetus are given by Cox.[94]

E. Other aspects

Considerable efforts have been devoted to modelling aspects of metabolism in the sheep using computers. Details are too complex to summarize here but details can be obtained for pool sizes and flows of labelled substances for sulphur,[119] phosphorus,[154] nitrogen,[282] magnesium,[257] thyroid hormones,[402] progesterone[289] and copper.[389]

Hepatic detoxification mechanisms include sulphate and glucuronide conjugation and possibly also phosphate conjugation.[213] The hepatic microsomal enzymes are very active and hence drugs such as pentobarbitone are metabolized rapidly.[115]

Most of the dietary choline is degraded in the rumen and only a small amount is synthetized by the liver. In spite of this, there are no signs of choline deficiency, probably as the oxidation rate of choline is low.[281]

The small intestine has been suggested as the main site of cholesterologenesis.[328] An increase in cholesterol synthesis in the small intestine following feeding of fat indicates that cholesterol may be associated in some way with fat absorption.[328]

3.11 Body Fluids and Renal Function

Body fluid compartments such as extracellular fluid and blood in theory should be relatively constant for any species when expressed on a body weight basis. However two factors in particular tend to cause the relative size per kg of the compartments in sheep to be variable. One of these is the volume of digesta in the gastro-intestinal tract as gut fill in sheep can be up to 40% of the body weight, an amount that is greater than in most monogastric animals. Hence variations in gut fill have a large influence on body weight. The second factor is the degree of fatness as, depending on previous feeding and management and the breed, the amount of carcase fat can range from negligible to about 50% or more of the carcase weight. A third factor which is not always taken into account when body weight of sheep is measured is the weight of wool carried. As a result of these factors, values for the body fluid compartments measured by different authors (Table 28) tend to be rather variable. Also the total body water and the plasma or blood volumes tend to be lower than for monogastric animals. Values cited for extracellular fluid volume in particular are variable for an additional reason that the marker substances which have been used do (thiosulphate, bromide) or do not (inulin, chromium EDTA) enter the gastro-intestinal tract and hence measure larger or smaller volumes.

Table 28 Body fluid volumes.

Parameter	Method	Value (ml/kg)
Total body water volume	^{3}HOH	460–533[242]
Total body water volume	Antipyrine	430–537[123]
Extracellular fluid volume	^{82}Br	245[85]
Extracellular fluid volume	Thiosulphate	200,[190] 149–155[123]
Red cell volume	^{59}Fe	21[380]
Plasma volume	T_{1824}	36·9,[254] 40–51,[123] 51·4[52]
Plasma volume	^{59}Fe	53[380]
Plasma volume	^{131}I-γ globulin	42·8,[85] 47[52]
Blood volume	^{59}Fe	53·9,[150] 74[380]
Blood volume	Calculated	56·2[85]
Blood volume	Calculated	64·9[52]

Total body water turnover under normal conditions is in the order of 10–14 ml/kg/day.[242] Under conditions of extreme heat where much water is needed for evaporative cooling, the turnover can increase considerably. As sheep are kept in hot arid regions in some countries, many studies have been done on mechanisms for withstanding heat and dehydration.[247] If a sheep in a temperate climate is deprived of water, eating ceases and the cessation of rumen function makes available most of the 4–5 litres of rumen water as well as water in the intestines for replenishment of normal water losses. As a result of this "stored" water, a sheep fasted in other than hot conditions does not need to drink for about 4 days.[177] Compared with man or other monogastric animals, sheep can lose much more body fluid without suffering permanent effects and recovery from losses of plasma volume and thiocyanate spaces of up to 45% have been recorded after 4 days without water in the sun in summer.[248] Such sheep when given water rapidly drank up to 10 litres.

In sheep fed intermittently (i.e. once or twice per day), the fluxes of water into and out of the gastro-intestinal tract in response to feeding are much greater than in monogastric animals. Not only is salivation profuse when dry foods are eaten but also the hypertonicity in the rumen caused by the production of low molecular weight compounds by fermentation also stimulates salivation.[388] In addition, there is movement of water across the rumen wall in response to an osmotic gradient.[387] The consequences of this reduction in plasma volume (to 20% [327]) and the associated metabolic acidosis caused by the secretion of large amounts of alkaline saliva are several. The haematocrit rises, partly due to the haemoconcentration and partly to a contraction of the spleen,[113] the systolic blood pressure rises abruptly and the heart rate rises more gradually but for a more prolonged period and there is renal conservation of sodium and water.[41] Unlike the "alkaline tide" seen in monogastric animals resultant from the secretion of gastric juice in response to feeding, there is a more acid urine formed in ruminants due to increased salivary bicarbonate secretion.[327]

Kidney function (Table 29) shows few peculiarities. The urine concentating power is greater than that of man, and sheep can survive by drinking 0·9% saline.[300] Unlike man, urate is actively secreted by the renal tubule.[77] Renal excretion of para-amino hippuric acid (PAH) is only 81–84% and hence estimates of renal blood flow made with this tracer are somewhat low unless corrections are made.[217] A creatinuria is normal.[42] A small paradoxical diuresis produced by infusing small amounts of antidiuretic hormone into non-diuresing sheep is caused by an increase in glomerular filtration induced by vascular changes in the kidney.[410]

Table 29 Kidney parameters.

Parameter	Value
Clearance PAH (ml/min/kg)	18·2,[126] 29·7[216]
Clearance PAH (ml/min)	321–381[415]
Clearance Inulin (ml/min/kg)	2·6[126]
Clearance Inulin (ml/min)	60–65[415]
Clearance Creatinine (ml/min/kg)	3·5[126]
Clearance Creatinine (ml/min)	82[190]
Clearance Thiosulphate (ml/min)	62[190]
Clearance Osmolality (ml/min)	1·06–1·44[415]
Glomerular filtration rate (ml/min)	118[218]
Glomerular filtration rate (ml/kg/min)	2·51[218]
T_m glucose (mg/min)	61–90[299]
T_m PAH (mg/min)	98[218]
T_m phosphate (μmol/min)	333[354]
T_m phosphate (μmol/100 ml glom. filtr.)	417[354]
T_m phosphate (μmol/$kg^{0·75}$)	19·2[354]
Maximum concentrating power (osmol/l)	3·8[249]

The T_m for phosphorus reabsorption by the kidney is higher for sheep than for most other animals that have been investigated. As a consequence, little phosphorus is normally excreted in the urine unless the intake is high. The major excretion route instead is the salivary glands with the phosphorus in excess of needs being excreted in the faeces.[355] This salivary phosphate which is important for buffering of rumen contents is under parathormone control.[353] As phosphate is normally unavailable for renal acid excretion, ammonia is used instead.[326]

The normal dietary intake of sheep is rich in potassium and low in sodium. Much of the potassium is excreted in faeces[176] or urine but some is in sweat. The watery extract of sweat is called suint and contains about 20% potassium on a dry weight basis.[343]

3.12 Climate Physiology

Every animal has a thermoneutral zone in which range of environmental temperature they do not expend energy to keep warm or keep cool. The thermoneutral zone is influenced by the recent history of the animal by such factors as the level of food intake and previous acclimatization. A factor of considerable importance in sheep is the fleece, the insulation of which reduces heat loss at all temperatures. As a guide, the thermoneutral zone of sheep is in the order of 26–28°C for a closely shorn sheep, 8–11°C for one with 4–4·5 cm of wool and 0–5°C for one with 10–12 cm of wool.[9] Thus the insulation value of the fleece causes a fully fleeced sheep to be relatively resistant to sub-zero temperatures. When the heat loss to the environment is greater than the resting metabolic heat production, sheep start shivering. Shearing produces an environmental stress and unless the weather is very warm a newly shorn sheep will shiver for several days.

Table 30 Relation between metabolic rate, coat thickness and critical temperature (from [44]).

State	Coat thickness (mm)	Metabolic rate (kcal/m^2/day)	Critical temperature (°C)
Maintenance ration fed	1	1250	28
Fasting	5	850	31
Maintenance ration fed	5	1250	25
Full ration fed	5	1600	18
Maintenance ration fed	10	1250	22
Maintenance ration fed	50	1250	9
Maintenance ration fed	100	1250	−3

(Critical temperature—environmental temperature below which metabolic rate increases to maintain homeostasis.)

The new born lamb is sensitive to cold shock and losses of lambs can be serious when lambs are born in cold wet windy weather. The new born lamb has a thermoneutral zone at about 25–30°C in which the metabolic heat production is 60–70 watts/m^2. At 0°C, this heat production is doubled in still air and tripled in a wind of 5·5 m/s.[6] The lamb has brown fat, rich in mitachondria, and stimulation of metabolism in this brown fat by adrenaline can rapidly increase the metabolic rate several fold. Even so, heat loss associated with a cold wet wind can exceed this maximal metabolic heat production. Once the lamb is dry and has drunk after birth, it is less prone to hypothermia. The brown fat of the lamb is concentrated in certain regions but all of the fat of the new born lamb has microscopic characteristics of brown fat. In the first 2–3 weeks after birth this is progressively converted to white adipose tissue.[149]

In cold environments, sheep exhibit two forms of cold-induced vasodilation (hunting phenomenon and continuous proportional control of skin temperature just above freezing point) and there appear to be individual differences as to which form a sheep employs.[268]

In sheep with short wool during winter, there is probably establishment of a lower thermoregulatory "set-point" and lower rectal temperatures.[392] The fleece also tends to keep sheep cool under conditions of high solar radiation. Temperatures as high as 85°C have been measured at the wool tip on the backs of Merino sheep in the sun. These sheep typically have a fleece length of 4 cm and a skin temperature of 42°C which indicates an insulation value of the fleece of greater than 10°C/cm.[247] Finer fleeces have greater insulation per cm

Table 31 Values for temperature, insulation and other data.

Parameter	Value
Body temperature (arterial) (°C)	39·06[57]
(rectal) (°C)	40·2,[163] 39·3[141]
(vaginal) (°C)	39·0–39·9[92]
Mean daily temperature variation (°C)	0·95[45]
Rise in rectal temperature with feeding (°C)	0·7[392]
Seasonal temperature variation (°C)	0·6–0·8[45]
Wool insulation (°C/m^2/W/mm)	0·015–0·023[6]
Tissue insulation (°C/W/m^2)	0·07–0·18[6]
Tissue insulation (°C/W/m^2)	1·2–3·5[44]
Thermoneutral zone (in water) (°C)	33–35[182]
Summit metabolism (W/kg)	10·07 (or 25 W/kg$^{0·75}$)[35]
Cutaneous evaporative heat loss (kcal/kg/day)	3–5 kcal/kg/day[57]
Respiratory evaporative heat loss (kcal/kg/day)	2–18[57]
Maximum sweating rate (g/m^2/h)	32[60]
Water turnover ($t_{1/2}$) (days)	3·35–4·25[247]

thickness but the hill breed sheep which may be exposed to cold environments have coarse but longer wool and the length compensates for the coarseness.

Most sheep have sweat glands (Table 32) although a few Merino sheep have a genetic absence of sweat glands. Lack of sweat glands is not a serious disadvantage as sweating is only a minor component of cooling.[60] The major cation in sweat is potassium rather than sodium and this is associated with the high intake of potassium by herbivores.

Table 32 Comparison of sweat gland function in sheep and man (from [60]).

	Sheep	Man
Gland density	290	150/cm
Volume of secretory part	0·004	0·003 mm^3
Secretion rate/gland	0·1	1·3 ml/h
Secretion rate/m^2	32	2000 g/h

The main source of evaporative cooling is panting (p. 38) where most of the evaporation is from the nasal epithelium. A small part of the fluid comes from the lateral nasal gland.[207] The secretion of this gland is probably typical of nasal secretion in being hypotonic (35–165 m osmol/litre) due to a reduced sodium content as compared with plasma.

The arterial supply to the brain is in close contact with venous blood in the carotid rete and it has been shown that the cooled nasal venous blood is able to cool the arterial blood passing to the brain.[412] This mechanism allows the brain to be kept at a temperature about 0·5°C lower than aortic blood at normal environmental temperatures and over 1·0°C in heat.[18]

The scrotum of the ram presents a special case as it has the function of keeping the testes at about 5°C below the rectal temperature.[141] Under cold conditions, the internal cremaster muscle contracts to draw the scrotum towards the abdomen to reduce its heat loss. The scrotum is liberally supplied with sweat glands and in heat these glands secrete more profusely than the glands over the remainder of the body. All sheep sweat glands are adrenergically innervated but one small difference between scrotal and other glands is that the scrotal glands respond to both adrenaline and noradrenaline whereas other sweat glands respond only to adrenaline.[308]

An interesting phenomenon is that, by heating the scrotum with a hair dryer, the scrotal rise in temperature causes the ram to pant[382] and this panting can cause sufficient cooling to induce shivering which may occur simultaneously with the panting.[392]

Putting pregnant ewes in very hot conditions induces foetal effects as lambs are born at about the normal time but at about two thirds of the normal weight and are very sparcely covered with wool.[69] The normal foetus is at a temperature of about 0·6°C greater than the maternal temperature (this is rarely recognized when foetal blood gas parameters are measured) but the differences decreases when the ewes are put in hot conditions.[72]

3.13 Pharmacology

Information on some half lives of drugs in sheep and responses of heart rate and blood pressure to some drugs is given in Tables 33 and 34 opposite. Other information is given in Table 35 on p. 92.

Table 33 Half lives of some drugs.

Drug	Half life
Pregnant Mares Serum	21·2 h[185]
25-hydroxycalciferol	3·1 weeks[251]
Lignocaine	31 min[274]
Bishydroxycoumarin	688 min[333]
Nitrate	0·475 h[324]
Nitrite	4·23 h[324]
Digoxin	0·5, 24[a] h[8]
Cyclophosphamide	2·6–2·8, 90–100[a] h[323]
Pentobarbitone	66·8 min[115]

[a] Second "compartment".

Table 34 Heart rate and mean arterial pressure responses to some commonly used drugs.

Drug	Dose	Heart rate	Mean arterial pressure
Atropine[405]	200 ug/kg/min	I 35–40%	
Atropine[406]	200 ug/kg/min	I 78–132%	0
Acetylcholine[406]	0·1–0·8 ug/kg	I 67%	D 13%
Angiotensin[283]	0·1 ug/kg	D 28%	U 53%
Isoproterenol[406]	0·2–0·4 ug/kg	I 160%	0
Methoxamine[283]	50 ug/kg/min	D 33%	I 55%
Nitroglycerine[283]	25 ug/kg	I 59%	D 25%
Noradrenaline[406]	0·1 ug/kg		I 20%
Phenoxybenzamine[406]	100 ug/kg	D 10–15%	0
Propanolol[283]	100–200 ug/kg	D 10–15%	0
Tyramine[283]	100 ug/kg	0	I 20%

I = increase; D = decrease; 0 = no change.

Table 35 Miscellaneous effects of drugs.

Drug	Comment
Propylene glycol	Commercial solutions of pentobarbitone sodium made up in a solution of propylene glycol can cause haemolysis.[298] Ethanol (10%) in water is preferred.
Serotonin and Histamine	The former contracts the pulmonary artery, trachea and bronchus but relaxes the pulmonary vein whereas the latter contracts the pulmonary artery and vein and trachea but relaxes the bronchi.[128] However there are dose effects.[129] Serotonin is thought not to be a natural mediator of the inflammatory reaction but acts as a histamine liberator.[376]
Insulin	1·0 *i.u.*/kg decreases blood glucose from 40 to 10 mg/ 100 ml.[390]
Adrenaline	0·75 mg/kg (subcutaneous, in oil) produces long-term increase in plasma glucose (to 400 mg/100 ml) and induces glucoseuria.[146] Five μg/kg (*i.v.*) produces cardiac irregularities but these can be eliminated by pretreatment with acepromazine.[306]
Chlorpromazine	Intravenous injections cause rapid increases in glucose and free fatty acids in blood due to adrenaline release.[62]
Atropine	Sheep salivary secretion is relatively insensitive to atropine. Scopolamine (0·03 mg/kg) may be a better inhibitor.[89]
Heparin	1 mg/kg (i/v) increases whole blood clotting time for 4·5–5 h while 2 mg/kg increases it for 5·5–6 h.[145]
Muscle relaxants	Sheep are more susceptible than man to curarine-type relaxants but are relatively resistant to gallamine.[71]
Prostaglandins	PGI_2 causes a dose related increase in pulmonary artery pressure.[51]

References

1. Abrahams, S. F., Coghlan, J. P., Denton, D. A., McDougall, J. G. and Mouw, D. R. (1976). Increased water drinking induced by sodium depletion in sheep. *Quart. J. exp. Physiol.*, **61**: 185–192.
2. Addonizio, V. P., Edmunds, L. H. and Colman, R. W. (1978). The function of monkey (*M mulatta*) platelets compared to platelets of pig, sheep and man. *J. lab. clin. Med.*, **91**: 989–997.
3. Agar, N. S., Evans, J. V. and Roberts, J. (1972). Red blood cell potassium and haemoglobin polymorphism in sheep: A review. *Anim. breeding Abstr.*, **40**: 407–436.
4. Agar, N. S. and Smith, J. E. (1973). Erythrocyte enzymes and glycolytic intermediates of high- and low-glutathione sheep. *Anim. Blood Grps. biochem. Genet.*, **4**: 133–140.
5. Akbar, A. M., Nett, T. M. and Niswender, G. D. (1974). Metabolic clearance and secretion rates of gonadotropins at different stages of the estrous cycle in ewes. *Endocrinology.*, **94**: 1318–1324.
6. Alexander, G. (1974). Heat loss from sheep: Assessment and control, pp. 173–203. In *Heat Loss from Animals and Man* (edited by Monteith, J. L. and Mount, L. E.). Butterworths; London.
7. Allsop, T. F. and Pauli, J. V. (1975). Responses to the lowering of magnesium and calcium concentrations in the cerebrospinal fluid of unanaesthetized sheep. *Aust. J. Biol. Sci.*, **28**: 475–481.
8. Antoniewicz, A. M., Heinemann, W. W. and Hanks, E. M. (1980). The effect of changes in the intestinal flow of nucleic acids on allantoin excretion in the urine of sheep. *J. agric. Sci.*, **95**: 395–400.
9. Armstrong, D. G., Blaxter, K. L., Graham, N. McC. and Wainman, F. W. (1959). The effect of environmental conditions on food utilization by sheep. *Anim. Prod.*, **1**: 1–12.
10. Armstrong, D. G. and Hutton, K. (1975). Fate of nitrogenous compounds entering the small intestine, pp. 432–447. (In reference 246.)
11. Arnold, G. W. (1970). Regulation of food intake in grazing ruminants, pp. 264–276. (In reference 297.)
12. Ash, R. W. (1961). Stimuli influencing the secretion of acid by the abomasum of sheep. *J. Physiol.*, **157**: 185–207.
13. Baber, K. A., Meyers, K. M., Clemmons, R. and Peters, R. (1979). Effects of tryptophan loading on the metabolism of serotonin in the central nervous system of the sheep. *Am. J. vet. Res.*, **40**: 1381–1385.
14. Baile, C. A. (1975). Control of feed intake in ruminants, pp. 330–350. (In reference 246.)
15. Baile, C. A. and Forbes, J. M. (1974). Control of feed intake and regulation of energy balance in ruminants. *Physiol. Rev.*, **54**: 160–214.
16. Baile, C. A. and Mayer, J. (1970). Hypothalamic centres: Feed-backs and receptor sites in the short-term control of feed intake, pp. 254–263. (In reference 297.)
17. Baird, D. T. (1978). Pulsatile secretion of LH and ovarian estradiol

during the follicular phase of the sheep estrous cycle. *Biol. Reprod.*, **18**: 359–364.

18. Baker, M. A. and Hayward, J. N. 1968). The influence of the nasal mucosa and the carotid rete upon hypothalamic temperature in sheep. *J. Physiol.*, **198**: 561–579.
19. Barta, O., Barta, V., Shirley, R. A. and McMurry, J. D. (1975). Haemolytic assay of sheep serum complement. *Zentrabl. vet. Med.*, **22B**: 254–262.
20. Barta, O. and Hubbert, N. L. (1978). Testing of hemolytic complement components in domestic animals. *Am. J. vet. Res.*, **39**: 1303–1308.
21. Bash, J. A. and Milgrom, F. (1972). Studies on allotypes in sheep. *Int. Arch. Allergy*, **42**: 196–214.
22. Bassett, J. M. (1974). Early changes in plasma insulin and growth hormone levels after feeding in lambs and adult sheep. *Aust. J. biol. Sci.*, **27**: 157–166.
23. Bassett, J. M. (1975). Dietary and gastro-intestinal control of hormones regulating carbohydrate metabolism in ruminants, pp. 383–398. (In reference 246.)
24. Bassett, J. M. (1978). Endocrine factors in the control of nutrient utilization: Ruminants. *Proc. Nutr. Soc.*, **37**: 273–280.
25. Bauchop, T. (1979). Rumen anaerobic fungi of cattle and sheep. *Appl. Environ. Microbiol.*, **38**: 148–158.
26. Bauman, D. E. and Davis, C. L. (1975). Regulation of lipid metabolism, pp. 496–509. (In reference 246.)
27. Baumann, R., Bauer, C. and Haller, E. A. (1975). Oxygen linked CO_2 transport in sheep blood. *Am. J. Physiol.*, **229**: 334–339.
28. Bayard, F., Louvet, J. P., Ruckebusch, Y. and Boulard, Cl. (1972). Transplacental passage of dexamethasone in sheep. *J. Endocr.*, **54**: 349–350.
29. Beal, A. M. and Bligh, J. (1977). Electrolyte concentrations in sheep cerebrospinal fluid. *Res. vet. Sci.*, **22**: 382–383.
30. Bedford, C. A., Harrison, F. A. and Heap, R. B. (1972). The metabolic clearance rate of progesterone and the conversion of progesterone to 20α-hydroxypregn-4-en-3-one in the sheep. *J. Endocr.*, **55**: 105–118.
31. Beh, K. J., Husband, A. J. and Lascelles, A. K. (1979). Intestinal response of sheep to intraperitoneal immunization. *Immunology*, **37**: 385–388.
32. Beisembaeva, R. U. and Albilova, G. M. (1978). Genetic polymorphism of sheep haptoglobin. *Genetika*, **14**: 1055–1058. (Cited in *Biol. Abstr.*, **67**: 35032.)
33. Benedictis, G. de, Capalbo, P. and Gallina, E. (1979). Immunogenetics of an antigen identified in both sheep and cattle sera. *Genet. Res.*, **34**: 41–46.
34. Ben-Ghedalia, D., Tagari, H., Bondi, E. and Tadmor, A. (1974). Protein digestion in the intestine of sheep. *Br. J. Nutr.*, **31**: 125–142.
35. Bennett, J. W. (1972). The maximum metabolic response of sheep to cold: Effects of rectal temperature, shearing, feed consumption, body posture, and body weight. *Aust. J. agric. Res.*, **23**: 1045–1058.

36. Bergland, R., Blume, H., Hamilton, A., Monica, P. and Patterson, R. (1980). Adrenocorticotropic hormone may be transported directly from the pituitary to the brain. *Science*, **210**: 541–543.
37. Bergman, E. N. and Heitmann, R. N. (1978). Metabolism of amino acids by the gut, liver, kidneys, and peripheral tissues. *Fed. Proc.*, **37**: 1228–1232.
38. Berman, W., Ravenscroft, P. J., Sheiner, L. B., Heymann, M. A., Melmon, K. L. and Rudolph, A. M. (1977). Differential effects of digoxin at comparable concentrations in tissues of fetal and adult sheep. *Circ. Res.*, **41**: 635–642.
39. Berman, W., Musselman, J. and Shortencarrier, R. (1980). The physiological effects of digoxin under steady-state drug conditions in newborn and adult sheep. *Circulation*, **62**: 1165–1171.
40. Bindon, B. M., Blanc, M. R., Pelletier, J., Terqui, M. and Thimonier, J. (1979). Periovulatory gonadotrophin and ovarian steriod patterns in sheep of breeds with differing fecundity. *J. Reprod. Fertil.*, **55**: 15–25.
41. Blair-West, J. R. and Brook, A. H. (1969). Circulatory changes and renin secretion in sheep in response to feeding. *J. Physiol.*, **204**: 15–30.
42. Blanch, E. and Setchell, B. P. (1960). Urinary excretion of creatine in the sheep. *Aust. J. biol. Sci.*, **13**: 356–360.
43. Bland, R. D., Demling, R. H., Selinger, S. L. and Staub, N. C. (1977). Effects of alveolar hypoxia on lung fluid and protein transport in unanesthetized sheep. *Circ. Res.*, **40**: 269–274.
44. Blaxter, K. L. (1962). *The Energy Metabolism of Ruminants*. Hutchinson; London.
45. Bligh, J., Ingram, D. L., Keynes, R. D. and Robinson, S. G. (1965). The deep body temperature of an unrestrained Welsh Mountain sheep recorded by a radiotelemetric technique during a 12 month period. *J. Physiol.*, **176**: 136–144.
46. Blunt, M. H. (1975). Cellular elements of ovine blood, pp. 29–44. (In reference 47.)
47. Blunt, M. H. (1975), editor. *The Blood of Sheep: Composition and Function*. Springer-Verlag; Berlin.
48. Blunt, M. H. and Huisman, T. H. J. (1975). The haemoglobins of sheep, pp. 155–183. (In reference 47.)
49. Borresen, A. L. (1977). High density lipoprotein (HDL) polymorphisms in rabbit: Production of antibody to rabbit allotype (R67) in sheep. *J. Immunogenetics*, **4**: 149–158.
50. Bouchard, R. and Brisson, G. J. (1969). Changes in protein fractions of ewe's milk throughout lactation. *Can. J. anim. Sci.*, **49**: 143–149.
51. Bowers, R. E., Ellis, E. F., Brigham, K. L. and Oates, J. A. (1979). Effects of prostaglandin cyclic endoperoxides on the lung circulation of unanesthetized sheep. *J. clin. Invest.*, **63**: 131–137.
52. Boyd, G. W. (1967). The reproducibility and accuracy of plasma volume estimation in the sheep with both ^{131}I Gamma Globulin and Evans blue. *Aust. J. exp. Biol. med. Sci.*, **45**: 51–75.

53. Bray, A. (1978). *Control of Luteal Life Span in Ewes*. Ph.D. Thesis, University of New England, Armidale, N.S.W., Australia.
54. Brockman, R. P. (1977). Glucose–glucagon relationships in adult sheep. *Can. J. comp. Med.*, **41**: 95–97.
55. Brockman, R. P. and Bergman, E. N. (1975). Qualitative aspects of insulin secretion and its hepatic and renal removal in sheep. *Am. J. Physiol.*, **229**: 1338–1343.
56. Brockman, R. P., Bergman, E. N., Pollak, W. L. and Brondum, J. (1975). Studies of glucose production in sheep using (6-^{3}H)glucose and (U-^{14}C)glucose. *Canad. J. Physiol. Pharmacol.*, **53**: 1186–1189.
57. Brockway, J. M., McDonald, J. D. and Pullar, J. D. (1965). Evaporative heat-loss mechanisms in sheep. *J. Physiol.*, **179**: 554–568.
58. Brockway, J. M. and McEwan, E. H. (1969). Oxygen uptake and cardiac performance in the sheep. *J. Physiol.*, **202**: 661–669.
59. Brook, A. H., Radford, H. M. and Stacy, B. D. (1968). The function of antidiuretic hormone in the sheep. *J. Physiol.*, **197**: 723–734.
60. Brook, A. H. and Short, B. F. (1960). Sweating in sheep. *Aust. J. agric. Res.*, **11**: 557–569.
61. Bruere, A. N. and Ellis, P. M. (1979). Cytogenetics and reproduction of sheep with multiple centric fusions (Robertsonian translocations). *J. Reprod. Fertil.*, **57**: 363–375.
62. Bruss, M. L. (1980). Effects of chlorpromazine on plasma concentrations of long chain fatty acids and glucose in sheep. *J. vet. Pharmacol.*, **3**: 35–44.
63. Bueno, L., Fioramonti, J. and Ruckebusch, Y. (1978). Postoperative intestinal motility in dogs and sheep. *Am. J. dig. Dis.*, **23**: 682–689.
64. Butcher, P. D. and Hawkey, C. M. (1979). The nature of erythrocyte sickling in sheep. *Comp. Biochem. Physiol.*, **64A**: 411–418.
65. Campbell, S. G., Siegel, M. J. and Knowlton, B. J. (1977). Sheep immunoglobulins and their transmission to the neonatal lamb. *N.Z. vet. J.*, **25**: 361–365.
66. Campbell, T. and Heath, T. (1973). Intrinsic contractility of lymphatics in sheep and in dogs. *Quart. J. exp. Physiol.*, **58**: 207–217.
67. Caple, I. and Heath, T. (1972). Regulation of output of electrolytes in bile and pancreatic juice in sheep. *Aust. J. biol. Sci.*, **25**: 155–165.
68. Caple, I. W. and Heath, T. J. (1975). Bilary and pancreatic secretions in sheep: Their regulation and roles, pp. 91–100. (In reference 246.)
69. Cartwright, G. A. and Thwaites, C. J. (1976). Foetal stunting in sheep: 1. The influence of maternal nutrition and high ambient temperature on the growth and proportions of Merino foetuses. 2. The effects of high ambient temperature during gestation on wool follicle development in the foetal lamb. *J. agric. Sci.*, **86**: 573–580, 581–585. (See also Brown, D. E., Harrison, P. C., Hinds, F. C., Lewis, J. A. and Wallace, M. H. (1977). *J. anim. Sci.*, **44**: 442–446.)
70. Carver, J. G. and Mott, J. C. (1978). Renin substrate in plasma of unanaesthetized pregnant ewes and their foetal lambs. *J. Physiol.*, **276**: 395–402.

71. Cass, N. M., Lampard, D. G., Brown, W. A. and Coles, J. R. (1976). Computer controlled muscle relaxation: A comparison of four muscle relaxants in the sheep. *Anaesth. Intens. Care*, **4**: 16–22.
72. Cefalo, R. C. and Hellegers, A. E. (1978). The effects of maternal hyperthermia on maternal and fetal cardiovascular and respiratory function. *Am. J. Obstet. Gynecol.*, **131**: 687–694.
73. Challis, J. R. G., Harrison, F. A. and Heap, R. B. (1973). The metabolic clearance rate, production rate and conversion ratios of oestrone in the sheep. *J. Endocr.*, **58**: 435–446.
74. Challis, J. R. G., Harrison, F. A. and Heap, R. B. (1973). The kinetics of oestradiol-17β metabolism in the sheep. *J. Endocr.*, **57**: 97–110.
75. Chamley, J. H. and Holland, R. A. B. (1969). Some respiratory properties of sheep haemoglobins A, B and C. *Respiration Physiol.*, **7** : 287–294.
76. Chamley, W. A., Stelmasiak, T. and Bryant, G. D. (1975). Plasma relaxin immunoactivity during the oestrous cycle of the ewe. *J. Reprod. Fertil.*, **45**: 455–461.
77. Chesley, L. C., Holm, L. W., Parker, H. R. and Assali, N. S. (1978). Renal tubular secretion of urate in sheep. *Proc. Soc. exp. Biol.*, **159**: 386–389.
78. Chopra, I. J., Sack, J. and Fisher, D. A. (1975). 3,3′,5′-Triiodothyronine (Reverse T_3) and 3,3′,5-Triiodothyronine (T_3) in fetal and adult sheep: Studies of metabolic clearance rates, production rates, serum binding, and thyroidal content relative to thyroxine. *Endocrinology*, **97**: 1080–1088.
79. Church, D. C. (1969). *Digestive Physiology and Nutrition of Ruminants. Vol. 1, Digestive Physiology*. O.S.V. Book Stores Inc., Corvallis, Oregon.
80. Clark, L. (1980). The electronic sheep. *Rural Research*, **No. 108**: 18–21.
81. Clarke I. J., Scaramuzzi, R. J. and Short, R. V. (1976). Effects of testosterone implants in pregnant ewes on their female offspring. *J. Embryol. exp. Morphol.*, **36**: 87–99.
82. Clarke, I. J., Scaramuzzi, R. J. and Short, R. V. (1977). Ovulation in prenatally androgenized ewes. *J. Endocr.*, **73**: 385–389.
83. Clarke, P. G. H. and Whitteridge, D. (1976). The cortical visual areas of the sheep. *J. Physiol.*, **256**: 497–508.
84. Coelho da Silva, J. F., Seeley, R. C., Thomson, D. J., Beaver, D. E. and Armstrong, D. G. (1972). The effect in sheep of physical form on the sites of digestion of a dried lucerne diet: 2 Sites of nitrogen digestion. *Br. J. Nutr.*, **28**: 43–61.
85. Coghlan, J. P., Fan, J. S. K., Scoggins, B. A. and Shulkes, A. A. (1977). Measurement of extracellular fluid volume and blood volume in sheep. *Aust. J. biol. Sci.*, **30**: 71–84.
86. Cole, G. J. and Morris, B. (1971). The growth and development of lambs thymectomized *in utero*. *Aust. J. exp. Biol. med. Sci.*, **49**: 33–53. (See also **49**: 55–73, 75–78 and 89–99.)
87. Cole, G. J. and Morris, B. (1973). The lymphoid apparatus of the sheep: its growth, development and significance in immunologic reactions. *Adv. vet. Sci. comp. Med.*, **17**: 225–263.

88. Coleman, G. S. (1975). The interrelationships between rumen ciliate protozoa and bacteria, pp. 149–164. (In reference 246.)
89. Collan, R. (1970). Anesthetic and paraoperative management of sheep for total heart replacement. *Anesth. Analg.*, **49**: 336–343.
90. Collis, S. C., Millson, G. C. and Kimberlin, R. H. (1977). Genetic markers in Herdwick sheep: No correlation with susceptibility or resistance to experimental scrapie. *Anim. blood Grps. biochem. Genet.*, **8**: 79–83.
91. Coombe, J. B. and Kay, R. N. B. (1965). Passage of digesta through the intestines of the sheep: Retention times in the small and large intestine. *Br. J. Nutr.*, **19**: 325–338.
92. Cooper, K. E., Kasting, N. W., Lederis, K. and Veale, W. L. (1979). Evidence supporting a role for endogenous vasopressin in natural suppression of fever in the sheep. *J. Physiol.*, **295**: 33–45.
93. Cornelius, C. E. and Gronwall, R. R. (1968). Congenital photosensitivity and hyperbilirubinemia in Southdown sheep in the United States. *Am. J. vet. Res.*, **29**: 291–295.
94. Cox, R. I. (1975). The endocrinologic changes of gestation and parturition in the sheep. *Adv. vet. Med. comp. Med.*, **19**: 287–305.
95. Crenshaw, C. and Cefalo, R. (1974). Effects of exogenous estrogen on P_{O_2} and experimental endotoxemia in sheep. *Am. J. Obstet. Gynec.*, **120**: 678–689.
96. Crighton, D. B. and Foster, J. P. (1977). Luteinizing hormone release after two injections of synthetic luteinizing hormone releasing hormone in the ewe. *J. Endocr.*, **72**: 59–67.
97. Cumming, I. A., Buckmaster, J. M., Blockey, M. A. deB., Goding, J. R., Winfield, C. G. and Baxter, R. W. (1973). Constancy of interval between luteinizing hormone release and ovulation in the ewe. *Biol. Reprod.*, **9**: 24–29.
98. Curtain, G. C. (1975). The ovine immune system, pp. 185–195. (In reference 47.)
99. Dabre, P. D., Adamson, J. W., Wood, W. G., Weatherall, D. J. and Robinson, J. S. (1979). Patterns of globin chain synthesis in erythroid colonies grown from sheep marrow of different development stages. *Br. J. Haematol.*, **41**: 459–475.
100. Dain, A. R. (1974). A study of the proportions of male and female leucocytes in the blood of chimaeric sheep. *J. Anat.*, **118**: 553–59.
101. Damon, E. G., Yelverton, J. T., Luft, U. C., Mitchell, K. and Jones, R. J. (1971). Acute effects of air blast on pulmonary function in dogs and sheep. *Aerospace Med.*, **42**: 1–9.
102. Davis, S. L. and Borger, M. L. (1973). Metabolic clearance rates and secretion rates of prolactin in sheep. *Endocrinology*, **92**: 1414–1418.
103. Davis, S. L. Ohlson, D. L., Klindt, J. and Anfinson, M. S. (1977). Episodic growth hormone secretory patterns in sheep: Relationship to gonadal steroid hormones. *Am. J. Physiol.*, **233**: E519–E523.
104. Davis, S. L., Ohlson, D. L., Klindt, J. and Anfinson, M. S. (1978). Episodic patterns of prolactin and thyrotropin secretion in rams and

wethers: Influence of testosterone and diethylstilbestrol. *J. anim. Sci.*, **46**: 1724–1729.
105. Davis, S. L., Ohlson, D. L., Klindt, J. and Everson, D. O. (1979). Estimates of repeatability in the temporal patterns of secretion of growth hormone (GH), prolactin (PRL) and thyrotropin (TSH) in sheep. *J. anim. Sci.*, **49**: 724–728.
106. Dawson, R. M. C. and Kemp, P. (1970). Biohydrogenation of dietary fats in ruminants, pp. 504–518. (In reference 297.)
107. Della-Fera, M. A. and Baile, C. A. (1979). Cholecystokinin octapeptide: continuous picomole injections into the cerebral ventricles of sheep suppress feeding. *Science*, **206**: 471–473.
108. Demeyer, D. I. and Van Nevel, C. J. (1975). Methanogenesis, an integrated part of carbohydrate fermentation and its control, pp. 366–382. (In reference 246.)
109. Denamur, R., Martinet, J. and Short, R. V. (1973). Pituitary control of the ovine corpus luteum. *J. Reprod. Fert.*, **32**: 207–220.
110. Dodds, W. J. (1978). Platelet function in animals: Species specificities, pp. 45–59. In *Platelets: a Multidisciplinary approach* (edited by Gaetano, G. de and Garattini, S). Raven Press; New York.
111. Dolling, C. E. and Good, B. F. (1976). Congenital goitre in sheep: Isolation of the iodoproteins which replace thyroglobulin. *J. Endocr.*, **71**: 179–192.
112. Domanski, E. and Polkoswska, J. (1973). The hypothalamic areas involved in the control of mammotrophic and lactotrophic processes in sheep. *Endocr. Experimentalis*, **7**: 229–246.
113. Dooley, P. C. (1973). *Contraction of the Sheep's Spleen*. Thesis for Doctor of Philosophy, University of New England, Armidale, Australia.
114. Dooley, P. C., Hecker, J. F. and Webster, M. E. D. (1972). Contraction of the sheep's spleen. *Aust. J. exp. Biol. med. Sci.*, **50**: 745–755.
115. Dos Santos, M. and Bogan, J. A. (1974). The metabolism of pentobarbitone in sheep. *Res. vet. Sci.*, **17**: 226–230.
116. Dougherty, R. W. (1965), editor. *Physiology of Digestion in the Ruminant*. Butterworths; London.
117. Dougherty, R. W., Allison, M. J. and Mullenax, C. H. (1964). Physiological disposition of C^{14}-labeled rumen gases in sheep and goats. *Am. J. Physiol.*, **207**: 1181–1188.
118. Dougherty, R. W., Hill, K. J., Cook, H. M. and Riley, J. L. (1971). Electromyographic and pressure studies of the oesophagus of the sheep. *Am. J. vet. Res.*, **32**: 1247–1252.
119. Doyle, P. T. and Moir, R. J. (1979). Sulfur and methionine metabolism in sheep: 1. First approximations of sulfur pools in and sulfur flows from the reticulo-rumen. *Aust. J. biol Sci.*, **32**: 51–63.
120. Edgerton, L. A. and Baile, C. A. (1977). Serum LH suppression by estradiol but not by testosterone or progesterone in wethers. *J. anim. Sci.*, **44**: 78–83.
121. Eiler, H., Lyke, W. A. and Johnson, R. (1981). Internal vomiting in the

ruminant: Effect of apomorphine on ruminal pH in sheep. *Am. J. vet. Res.*, **42**: 202–204.
122. Ellinwood, W. E., Nett, T. M. and Niswender, G. D. (1979). Maintenance of the corpus luteum of early pregnancy in the ewe: 2. Prostaglandin secretion by the endometrium *in vitro* and *in vivo*. *Biol. Reprod.*, **21**: 845–856.
123. Engelhardt, W. U. and Hauffe, R. (1975). Role of the omasum in absorption and secretion of water and electrolytes in sheep and goats, pp. 216–230. (In reference 246.)
124. English, P. B. (1966). A study of water and electrolyte metabolism in sheep: 2. The volumes of distribution of antipyrine, thiosulphate and T1824 (Evans blue) and values for certain extracellular fluid constituents. *Res. vet. Sci.*, **7**: 258–275.
125. English, P. B., Hardy, L. N. and Holmes, E. M. (1969). Values for plasma electrolytes, osmolality and creatinine and venous P_{CO_2} in normal sheep. *Am. J. vet. Res.*, **30**: 1967–1973.
126. English, P. B., Hogan, A. E. and McDougall, H. L. (1977). Changes in renal function with reductions in renal mass. *Am. J. vet. Res.*, **38**: 1317–1322.
127. Evans, H. E., Ingalls, T. H. and Binns, W. (1966). Teratogenesis of craniofacial malformations in animals: 3. Natural and experimental cephalic deformities in sheep. *Arch. environ. Health*, **13**: 706–714.
128. Eyre, P. (1975). Comparisons of reactivity of pulmonary vascular and airway smooth muscle of sheep and calf to tryptamine analogues, histamine and antigen. *Am. J. vet. Res.*, **36**: 1081–1084.
129. Eyre, P. (1975). Atypical tryptamine receptors in sheep pulmonary vein. *Br. J. Pharmacol.*, **55**: 329–333.
130. Eyre, P. and Burka, J. F. (1978). Hypersensitivity in cattle and sheep: A pharmacological review. *J. vet. Pharmacol. Therap.*, **1**: 97–109.
131. Fahey, K. J. and Brandon, M. R. (1978). Synthesis of immunoglobulin by normal and antigenically stimulated fetal sheep. *Res. vet. Sci.*, **25**: 218–224.
132. Fahey, K. J. and Morris, B. (1978). Humoral immune responses in foetal sheep. *Immunology*, **35**: 651–661.
133. Fantl, P. and Ward, H. A. (1960). Clotting activity of maternal and foetal sheep blood. *J. Physiol.*, **150**: 607–620.
134. Farrell, D. J., Leng, R. A. and Corbett, J. L. (1972). Undernutrition in grazing sheep: 1. Changes in the composition of the body, blood and rumen contents. 2. Calormetric measurements on sheep taken from pasture. *Aust. J. agric. Res.*, **23**: 483–497, 499–509.
135. Feinstein, A. and Hobart, M. J. (1969). Structural relationship and complement fixing activity of sheep and other ruminant immuglobulin G subclasses. *Nature*, **223**: 950–952.
136. Ferguson, K. A. and Cox, R. I. (1975). Hormones, pp. 101–121. (In reference 47.)
137. Fisher, D. A., Dussault, J. H., Sack, J. and Chopra, I. J. (1977).

Otogenesis of hypothalamic-pituitary-thyroid function and metabolism in man, sheep and rat. *Rec. progr. Hormone Res.*, **33**: 59–107.

138. Flint, A. P. F., Forsling, M. L. and Mitchell, M. D. (1978). Blockade of the Ferguson reflex by lumbar epidural anaesthesia in the parturient sheep: Effects on oxytocin secretion and uterine venous prostaglandin F levels. *Horm. Metab. Res.*, **10**: 545–547.

139. Ford, C. H. J. (1975). Genetic studies of sheep leucocyte antigens. *J. Immunogenetics*, **2**: 31–40.

140. Ford, C. H. J. and Elves, M. W. (1974). The production of cytotoxic antileucocyte antibodies by parous sheep. *J. Immunogenetics*, **1**: 259–264.

141. Free, M. J. and Van Demark, N. L. (1968). Gas tensions in spermatic and peripheral blood of rams with normal and heat-treated testes. *Am. J. Physiol.*, **214**: 863–865.

142. Fulkerson, W. J. (1978). Synchronous episodic release of cortisol in the sheep. *J. Endocr.*, **79**: 131–132.

143. Fulkerson, W. J., McDowell, G., Hooley, R. D. and Fell, L. R. (1977). Artificial induction of lactation in ewes: The use of prostaglandin. *Aust. J. biol. Sci.*, **30**: 573–576.

144. Fulkerson, W. J. and Tang, B. Y. (1979). Ultradian and circadian rhythms in the plasma concentrations of cortisol in sheep. *J. Endocr.*, **81**: 135–141.

145. Gajewski, J. and Povar, M. L. (1971). Blood coagulation values of sheep. *Am. J. vet. Res.*, **32**: 405–409.

146. Gans, J. H. and Biggs, D. L. (1968). Metabolic effects of long-acting epinephrine in sheep. *Am. J. vet. Res.*, **29**: 1167–1172.

147. Garel, J. M., Care, A. D. and Barlet, J. P. (1974). A radioimmunoassay for ovine calcitonin: An evaluation of calcitonin secretion during gestation, lactation and foetal life. *J. Endocr.*, **62**: 497–509.

148. Gattinoni, L. and Samaja, M. (1979). Acid-base equilibrium in the blood of sheep. *Experimentia*, **35**: 1347–1348.

149. Gemmell, R. T., Bell, A. W. and Alexander, G. (1972). Morphology, of adipose cells in lambs at birth and during subsequent transition of brown to white adipose tissue in cold and in warm conditions. *Am. J. Anat.*, **133**: 143–146.

150. Giles, R. C., Berman, A., Hildebrandt, P. K. and McCaffrey, R. P. (1977). Use of ^{59}Fe for sheep erythrocyte kinetic studies. *Am. J. vet. Res.*, **38**: 534–537.

151. Goodman, R. L., Pickover, S. M. and Karsch, F. J. (1981). Ovarian feedback control of follicle-stimulating hormone in the ewe: Evidence for selective suppression. *Endocrinology*, **108**: 772–777.

152. Gorrill, A. D. L., Schingoethe, D. J. and Thomas, J. W. (1968). Proteolytic activity and *in vitro* enzyme stability in small intestinal contents from ruminants and non-ruminants at different ages. *J. Nutr.*, **96**: 342–348.

153. Grabowski, E. F., Didisheim, P., Lewis, J. C., Franta, J. T. and Stropp, J. Q. (1977). Platelet adhesion to foreign surfaces under controlled conditions of whole blood flow: Human vs rabbit, dog, calf, sheep, pig,

macaque, and baboon. *Trans. Am. Soc. artif. internal Organs*, **23**: 141–149.

154. Grace, N. D. (1981). Phosphorus kinetics in the sheep. *Br. J. Nutr.*, **45**: 367–374.
155. Graham, W. F., Campbell, D. J., Coghlan, J. P. *et al.* (1980). Changes in the relationship between cerebrospinal fluid and blood composition produced by ACTH treatment in conscious sheep. *Life Sci.*, **26**: 2265–2271.
156. Green, D. and Moor, R. M. (1977). The influence of anaesthesia on the concentrations of progesterone and cortisol in peripheral blood plasma of sheep. *Res. vet. Sci.*, **22**: 122–123.
157. Greenwood, B. (1977). Haematology of the sheep and goat, pp. 305–344. In *Comparative Clinical Haematology* (edited by Archer, R. K. and Jeffcott, L. B.). Blackwell; Oxford.
158. Grivel, M-L. and Ruckebusch, Y. (1972). The propagation of segmental contractions along the small intestine. *J. Physiol.*, **227**: 611–625.
159. Grovum, W. L. and Hecker, J. F. (1973). Rate of passage of digesta in sheep: 2. The effect of level of food intake on digesta retention times and on water and electrolyte absorption in the large intestine. *Br. J. Nutr.*, **30**: 221–230.
160. Hales, J. R. S. (1973). Effects of exposure to hot environments on the regional distribution of blood flow and on cardiorespiratory function in sheep. *Pflugers Arch.*, **344**: 133–148.
161. Hales, J. R. S. and Brown, G. D. (1974). Net energetic and thermoregulatory efficiency during panting in the sheep. *Comp. Biochem. Physiol.*, **49A**: 413–422.
162. Hales, J. R. S., Dampney, R. A. L. and Bennett, J. W. (1975). Influences of chronic denervation of the carotid bifurcation regions on panting in the sheep. *Pflugers Arch.*, **360**: 243–253.
163. Hales, J. R. S. and Webster, M. E. D. (1967). Respiratory function during thermal tachypnoea in sheep. *J. Physiol.*, **190**: 241–260.
164. Hall, J. G. (1971). The lymph-borne cells of the immune response: A review, pp. 39–57. In *The Scientific Basis of Medicine Annual Reviews* (edited by Gilliland, I. and Francis, J.). The Athlone Press; University of London.
165. Halmagi, D. F. J. and Gillett, D. J. (1966). Cardiorespiratory consequences of corrected gradual severe blood loss in unanaesthetized sheep. *J. appl. Physiol.*, **21**: 589–596.
166. Hamlin, R. L., Smith, C. R. and Redding, R. W. (1960). Time-order of ventricular activation for premature beats in sheep and dogs. *Am. J. Physiol.*, **198**: 315–321.
167. Hammerberg, B., Brett, I. and Kitchen, H. (1977). Otogeny of hemoglobins in sheep. *Ann. N.Y. Acad. Sci.*, **241**: 672–681.
168. Handwerger, S., Crenshaw, C., Maurer, W. F., Barrett, J., Hurley, T. W., Golander, A. and Fellows, R. E. (1977). Studies on ovine placental lactogen secretion by homologous radioimmunoassay. *J. Endocr.*, **72** 27–34.
169. Harding, R., Johnson, P. and McClelland, M. E. (1978). Liquid-sensitive laryngeal receptors in the developing sheep, cat and monkey. *J. Physiol.*, **277**: 409–422.

170. Harrison, F. A. and Leat, W. M. F. (1975). Digestion and absorption of lipids in non-ruminant and ruminant animals: A comparison. *Proc. nutr. Soc.*, **34**: 203–210.
171. Hays, F. L. and Webster, A. J. F. (1971). Effects of cold, eating, efferent nerve stimulation and angiotensin on heart rate in sheep before and after autonomic blockade. *J. Physiol.*, **216**: 21–38.
172. Hazelwood, J. C. and Heath, G. E. (1976). A comparison of cholinesterase activity of plasma, erythrocytes, and cerebrospinal fluid of sheep, calves, dogs, swine and rabbits. *Am. J. vet. Res.*, **37**: 741–743.
173. Hearnshaw, H., Cummings, I. A. and Goding, J. R. (1977). Concentrations of luteinizing hormone in ovariectomized ewes and the effect of disturbance, starvation and vascular cannulation. *Aust. J. biol. Sci.*, **30:** 97–101.
174. Hecker, J. F. (1974). See reference 19 in Chapter 1.
175. Hecker, J. F. (1971). Ammonia in the large intestine of herbivores. *Br. J. Nutr.*, **26**: 135–145.
176. Hecker, J. F. and Grovum, W. L. (1971). Absorption of water and electrolytes from the large intestine of sheep. *Aust. J. biol. Sci.*, **24**: 365–372.
177. Hecker, J. F., Budtz-Olsen, O. E. and Ostwald, M. (1964). The rumen as a water store in sheep. *Aust. J. agric. Res.*, **15**: 961–968.
178. Hecker, J. F., Strakosch, T., Wodzicka-Tomaszewska, M. and Bray, A. R. (1974). The effects of size of intrauterine devices on luteal function in ewes. *Biol. Reprod.*, **11**: 73–78.
179. Heitmann, R. N. and Bergman, E. N. (1980). Transport of amino acids in whole blood and plasma of sheep. *Am. J. Physiol.*, **239**: E242–E247.
180. Heitmann, R. N. and Bergman, E. N. (1980). Integration of amino acid metabolism in sheep: Effects of fasting and acidosis. *Am. J. Physiol.*, **239**:E248–E254.
181. Hembry, F. G., Bell, M. C. and Hall, R. F. (1967). Intestinal carbohydrase activity and carbohydrate utilization in mature sheep. *J. Nutr.*, **93**: 175–181.
182. Hemingway, A. and Hemingway, C. (1966). Respiration of sheep at thermoneutral temperature. *Resp. Physiol.*, **1**: 30–37.
183. Hensby, C. N. (1974). Reduction of prostaglandin E_2 to prostaglandin $F_{2\alpha}$ by an enzyme in sheep blood. *Br. J. Pharmacol.*, **50**: 462P.
184. Herriman, I. D., Heitzman, R. J., Priestly, I. and Sandhu, G. S. (1976). Concentrations of intermediate metabolites in the blood and hepatic tissues of pregnant and non-pregnant ewes. *J. agric. Sci.*, **87**: 407–411. (See also *J. agric. Sci.*, **90**: 579–585 (1978)).
185. Hidiroglou, M., Williams, C. J. and Ivan, M. (1979). Pharmacokinetics and amounts of 25-hydroxycalciferol in sheep affected by osteodystrophy. *J. Dairy Sci.*, **62**: 567–571.
186. Hodgetts, V. E. (1961). The dynamic red cell storage function of the spleen in sheep: 3 Relationship to determination of blood volume, total red cell volume and plasma volume. *Aust. J. exp. Biol. med. Sci.*, **39**: 187–196.
187. Hofman, W. F. and Riegle, G. D. (1977). Thermorespiratory responses of shorn and unshorn sheep to mild heat stress. *Resp. Physiol.*, **30**: 327–338.

188. Hofmeyr, C. F. B. (1978). Evaluation of neurological examination of sheep. *J. S. Afri. vet. Assn.*, **49**: 45–48.
189. Holley, D. C., Beckman, D. A. and Evans, J. W. (1975). Effect of confinement on the circadian rhythm of ovine cortisol. *J. Endocr.*, **65**: 147–148.
190. Holmes, E. M. and English, P. B. (1969). The volume of distribution of sodium thiosulphate in sheep. *Res. vet. Sci.*, **10**: 73–78.
191. Holst, P. J. (1974). The time of entry of ova into the uterus of the ewe. *J. Reprod. Fertil.*, **36**: 427–428.
192. Hoogenraad, N. J. and Hird, F. J. R. (1970). The chemical composition of rumen bacteria and cell walls from rumen bacteria. *Br. J. Nutr.*, **24**: 119–127.
193. Hoogenraad, N. J., Hird, F. J. R., Holmes, I. and Mills, N. F. (1967). Bacteriophages in rumen contents of sheep. *J. gen. Virol.*, **1**: 575–576.
194. Hooper. R. H. and Short, A. H. (1977). The hepatocellular uptake of glucose, galactose and fructose in conscious sheep. *J. Physiol.*, **264**: 523–539.
195. Hoover, W. H., Young, P. J., Sawyer, M. S. and Apgar, W. P. (1970). Ovine physiological responses to elevated ambient carbon dioxide. *J. appl. Physiol.*, **29**: 32–35.
196. Hopkins, P. S., Wallace, A. L. C. and Thorburn, G. D. (1975). Thyrotrophin concentrations in the plasma of cattle, sheep and foetal lambs as measured by radioimmunoassay. *J. Endocr.*, **64**: 371–387.
197. Horino, M., Machlin, L. J., Hertelendy, F. and Kipnis, D. M. (1968). Effect of short-chain fatty acids on plasma insulin in ruminant and non ruminant species. *Endocrinology*, **83**: 118–128.
198. Huisman, T. H. J. and Kitchens, J. (1968). Oxygen equilibria studies of the hemoglobins from normal and anaemic sheep and goats. *Am. J. Physiol.*, **215**: 140–146.
199. Hungate, R. E. (1966). *The Rumen and its Microbes*. Academic Press; New York and London.
200. Iannelli, D. (1979). Immunogenetics of the A_2 serum antigen of sheep (*Ovis aries*). *Immunogenetics*, **8**: 77–81.
201. Ismay, M. J. A., Lumbers, E. R. and Stevens, A. D. (1979). The action of angiotensin II on the baroflex of the conscious ewe and the conscious foetus. *J. Physiol.*, **288**: 467–479.
202. Jaffe, B. M., Peskin, G. W. and Kaplan, E. L. (1974). Relationship of serum gastrin to parathyroid hormone secretion in sheep. *Metabolism*, **23**: 307–310.
203. Jarrett, I. G. (1972). A polymeric form of haemoglobin-binding protein in sheep following metabolic and hormonal disturbance. *Aust. J. biol. Sci.*, **25**: 941–948.
204. Jarrett, I. G., Potter, B. J. and Packham, A. (1956). Effects of pancreatectomy in the sheep. *Aust. J. exp. Biol.*, **34**: 133–142.
205. John, M. E. (1979). Sheep hemoglobin E: A third *Beta* chain allele. *Hemoglobin*, **3**: 93.

206. Johnson, J. I., Rubel, E. W. and Hatton, G. I. (1974). Mechanosensory projections to cerebral cortex of sheep. *J. comp. Neurol.*, **158**: 81–107.
207. Johnson, K. G. and Peaker, M. (1974). Studies of the lateral nasal (Steno's) glands of the sheep. *J. Physiol.*, **242**: 7P–8P.
208. Johnston, C. I. (1972). Radioimmunoassay for plasma antidiuretic hormone. *J. Endocr.*, **52**: 69–78.
209. Jones, C. T., Luther, E., Ritchie, J. W. K. and Worthington, D. (1975). The clearance of ACTH from the plasma of adult and fetal sheep. *Endocrinology.*, **96**: 231–234.
210. Kammula, R. G. (1976). Metabolism of ketone bodies by ovine brain *in vivo*. *Am. J. Physiol.*, **231**: 1490–1494.
211. Kaneko, J. J. (1974). Comparative erythrocyte metabolism. *Adv. vet. Med. comp. Med.*, **18**: 117–153.
212. Kaneko, J. J. and Cornelius, C. E. (1970), editors. *Clinical Biochemistry of Domestic Animals*, 2nd edition. Academic Press; New York and London.
213. Kao, J., Bridges, J. W. and Faulkner, J. K. (1979). Metabolism of ^{14}C phenol by sheep, pig and rat. *Xenobiotica*, **9**: 141–147.
214. Kaplan, A. M. and Freeman, M. J. (1972). Synthesis and properties of antiprotein antibodies of sheep: Primary antibody response. Secondary antibody response. *Am. J. vet. Res.*, **33**: 1947–1954, 1955–1961.
215. Karamanlidis, A. N. and Magras, J. (1972). Retinal projections in domestic ungulates: 1. The retinal projections in the sheep and the pig. *Brain Res.*, **44**: 127–145.
216. Katongole, C. B., Naftolin, F. and Short, R. V. (1974). Seasonal variations in blood luteinizing hormone and testosterone levels in rams. *J. Endocr.*, **60**: 101–106.
217. Kaufman, C. F. and Bergman, E. N. (1971). Cannulation of renal veins of sheep for long-term functional and metabolic studies. *Am. J. vet. Res.*, **32**: 1103–1107.
218. Kaufman, C. F. and Bergman, E. N. (1978). Renal function studies in normal and toxemic pregnant sheep. *Cornell Vet.*, **68**: 124–137.
219. Kay, R. N. B. (1960). The rate of flow and composition of various salivary secretions in sheep and calves. *J. Physiol.*, **150**: 515–537.
220. Keirse, M. J. N. C., Hicks, B. R. and Turnbull, A. C. (1975). Comparison of intra-uterine metabolism of prostaglandin $F_{2\alpha}$ in ovine and human pregnancy. *J. Endocr.*, **67**: 24P–25P.
221. Keller, P. (1973). The activity of enzymes in serum and tissues of clinically normal sheep. *N.Z. vet. J.*, **21**: 221–227.
222. Killeen, I. D. and Moore, N. W. (1970). Fertilization and survival of fertilized eggs in the ewe following surgical insemination at various times after the onset of oestrus. *Aust. J. biol. Sci.*, **23**: 1279–1287.
223. Kirk, E. J. (1968). The dermatomes of the sheep. *J. comp. Neurol.*, **134**: 353.
224. Klindt, J., Davis, S. L. and Ohlson, D. L. (1979). Estimation of half-life and metabolic clearance rate of thyrotrophin releasing hormone in sheep using a double antibody radioimmunoassay. *J. anim. Sci.*, **48**: 1165–1171.

225. Kung, M., Reinhart, M. E. and Wanner, A. (1978). Pulmonary hemodynamic effects of lung inflation and graded hypoxia in conscious sheep. *J. appl. Physiol.*, **45**: 949–956.
226. Land, R. B., Pelletier, J., Thimonier, J. and Mauleon, P. (1973). A quantitative study of genetic differences in the incidence of oestrus, ovulation and plasma luteinizing hormone concentration in the sheep. *J. Endocr.*, **58**: 305–317.
227. Landa, J. F., Hirsch, J. A. and Lebeaux, M. I. (1975). Effects of topical and general anesthetic agents on tracheal mucous velocity of sheep. *J. appl. Physiol.*, **38**: 946–948.
228. Lanz, E., Ruiz, B. and Schlereth, H. (1979). Local anesthetics and pulmonary venous admixture in pregnant and nonpregnant sheep. *Anesth. Analg.*, **58**: 318–321.
229. Laster, D. B. and Glimp, H. A. (1972). A note on the effect of ram to ewe ratio on reproductive performance of synchronized ewes. *Anim. Prod.*, **15**: 99–102.
230. Leat, W. M. F. (1970). Carbohydrate and lipid metabolism in the ruminant during post-natal development, pp. 211–222. (In reference 297.)
231. Leat, W. M. F. and Harrison, F. A. (1975). Digestion, absorption and transport of lipids in the sheep, pp. 481–495. (In reference 246.)
232. Leat, W. M. F., Kubasek, F. O. T. and Buttress, N. (1976). Plasma lipoproteins of lambs and sheep. *Quart. J. exp. Physiol.*, **61**: 193–202.
233. Lee, W. B., Ismay, M. J. and Lumbers, E. R. (1980). Mechanisms by which angiotensin II affects the heart rate of the conscious sheep. *Circ. Res.*, **47**: 286–292.
234. Lewis, D. (1961), editor. *Digestive Physiology and Nutrition of the Ruminant*. Butterworths; London.
235. Lieb, S. M., Cabalum, T., Zugaib, M. *et al.* (1980). Vascular reactivity to angiotensin II in the normotensive and hypertensive pregnant ewe. *Am. J. Physiol.*, **238**: H209–H213.
236. Liggins, G. C., Fairclough, R. J., Grieves, S. A., Kendall, J. Z. and Knox, B. S. (1973). The mechanisms of induction of parturition in the ewe. *Rec. progr. Hormone Res.*, **29**: 111–150.
237. Lincoln, G. (1979). Use of a pulsed infusion of luteinizing hormone-releasing hormone to mimic seasonally induced endocrine changes in the ram. *J. Endocr.*, **83**: 251–260.
238. Lindsay, D. B. (1970). Carbohydrate metabolism in ruminants, pp. 438–451. (In reference 297.)
239. Lindsay, D. B. and Leat, W. M. F. (1975). Carbohydrate and lipid metabolism, pp. 45–62. (In reference 47.)
240. Lindsay, D. B. and Setchell, B. P. (1976). The oxidation of glucose, ketone bodies and acetate by the brain of normal and ketonaemic sheep. *J. Physiol.*, **259**: 801–823.
241. Long, S. E. and Williams, C. V. (1980). Frequency of chromosomal abnormalities in early embryos of the domestic sheep (*Ovis aries*). *J. Reprod. Fertil.*, **58**: 197–201.

242. Longhurst, W. M., Baker, N. F., Connolly, G. E. and Fisk, R. A. (1970). Total body water and water turnover in sheep and deer. *Am. J. vet. Res.*, **31**: 673–677.
243. Lopaciuk, S., McDonagh, R. P. and McDonagh, J. (1978). Comparative studies on blood coagulation Factor XIII. *Proc. Soc. exp. Biol. Med.*, **158**: 68–72.
244. Lough, A. K. (1970). Aspects of lipid digestion in the ruminant, pp. 519–528. (In reference 297.)
245. Lun, S., Espiner, E. A. and Hart, D. S. (1978). Adrenocortical metabolism of angiotension in sheep with adrenal transplants. *Am. J. Physiol.*, **235**: E525–E531.
246. McDonald, I. W. and Warner, A. C. I. (1975), editors. *Digestion and Metabolism in the Ruminant.* University of New England Publishing Unit; Armidale, Australia.
247. MacFarlane, W. V. (1964). Terrestrial animals in dry heat: ungulates, pp. 509–538. In *Handbook of Physiology: Adaptations to the Environment.* American Physiological Society; Washington.
248. MacFarlane, W. V., Morris, R. J. and Howard, B. (1956). Water economy of tropical Merino sheep. *Nature.*, **178**: 304–305.
249. MacFarlane, W. V., Morris, R. J. H., Howard, B., McDonald, J. and Budtz-Olsen, O. E. (1961). Water and electrolyte changes in tropical Merino sheep exposed to dehydration during summer. *Aust. J. agric. Res.*, **12**: 889–912.
250. McIntosh, G. H., Filsell, O. H. and Jarrett, I. G. (1973). Kidney function and net glucose production in normal and acidotic sheep. *Aust. J. biol. Sci.*, **26**: 1389–1394.
251. McIntosh, J. E. A., Moor, R. M. and Allen, W. R. (1975). Pregnant mare serum gonadotrophin: Rate of clearance from the circulation of sheep. *J. reprod. Fertil.*, **44**: 95–100.
252. McKenzie, J. S. and Denton, D. A. (1974). Salt ingestion responses to diencephalic electrical stimulation in the unrestrained conscious sheep. *Brain Res.*, **70**: 449–466.
253. McKenzie, J. S. and Smith, M. H. (1973). Stereotaxic method and variability data for the brain of the Merino sheep. *J. fur Hirnforschung*, **14**: 355–366.
254. Mackie, W. S. (1976). Plasma volume measurements in sheep using Evan's blue and continuous blood sampling. *Res. vet. Sci.*, **21**: 108–109.
255. McNatty, K. P., Revfeim, K. J. A. and Young, A. (1973). Peripheral plasma progesterone concentrations in sheep during the oestrous cycle. *J. Endocr.*, **58**: 219–225.
256. McNatty, K. P. and Thurley, D. C. (1973). The episodic nature of changes in ovine plasma cortisol levels and their response to adrenaline during adaption to a new environment. *J. Endocr.*, **59**: 171–180.
257. Madsen, F. C., Spalding, G. E., Miller, J. K., Hansard, S. L. and Lyke, W. A. (1975). Magnesium movement in hypothyroid sheep. *Proc. Soc. exp. Biol. Med.*, **149**: 207–214.

258. Maloney, J. E., Alcorn, D., Cannata, J., Walker, A. and Ritchie, B. C. (1974). Regional vascular response to hypoxia in the lungs of anaesthetized sheep. *Aust. J. exp. Biol. med. Sci.*, **52**: 801–812.
259. Malven, P. V. and Mollett, T. A. (1978). Relationship between tachycardia, prolactin and growth hormone in conscious ewes. *Neuroendocrinology*, **27**: 160–174.
260. Manders, W. T., Pagani, M. and Vatner, S. F. (1979). Depressed responsiveness to vasoconstrictor and dilator agents and baroreflex sensitivity in conscious, newborn lambs. *Circulation*, **60**: 945–955.
261. Manwell, C. and Baker, C. M. A. (1977). Genetic distance between the Australian Merino and the Poll Dorset sheep. *Genet. Res.*, **29**: 239–253.
262. Margni, R. A., Castrelos, O. D. and Paz, C. B. (1973). The sheep immune response: Variation of anti-hapten and anti-carrier antibodies in the γ_1 and γ_2 immunoglobulin fractions. *Immunology*, **24**: 781–789.
263. Martal, J., Lacroix, M-C., Loudes, C., Saunier, M. and Wintenberger-Torres, S. (1979). Trophoblastin, an antiluteolytic protein present in early pregnancy in sheep. *J. Reprod. Fertil.*, **56**: 63–73.
264. Martin, G. B., Scaramuzzi, R. J., Cox, R. I. and Gherardi, P. B. (1979). Effects of active immunization against androstenedione or oestrone on oestrus, ovulation and lambing in Merino ewes. *Aust. J. exp. Agric. anim. Husb.*, **19**: 673–678.
265. Matthews, C. D., Kennaway, D. J., Frith, R. G., Phillipou, G., Le Cornu, A. and Seamark, R. F. (1977). Plasma melatonin values in man and some domestic animals: Initial observations on the effects of pregnancy in man and pinealectomy in sheep. *J. Endocr.*, **73**: 41P–42P.
266. Mattner, P. E. and Braden, A. W. H. (1967). Studies in flock mating of sheep: 2 Fertilization and prenatal mortality. *Aust. J. exp. Agric. anim. Husb.*, **7**: 110–116.
267. Mendel, V. E. and Raghavan, G. V. (1964). A study of diurnal temperature patterns in sheep. *J. Physiol.*, **174**: 206–216.
268. Meyer, A. A. and Webster, A. J. F. (1971). Cold-induced vasodilatation in the sheep. *Can. J. Physiol. Pharmacol.*, **49**: 901–908.
269. Millot, P. (1979). Genetic control of lymphocyte antigens in sheep: The OLA complex and two minor loci. *Immunogenetics*, **9**: 509–534.
270. Mitchell, B. and Williams, J. T. (1975). Normal blood-gas values in lambs during neonatal development and in adult sheep. *Res. vet. Sci.*, **19**: 335–336.
271. Miyazaki, T., Peric-Golia, L., Slaunwhite, W. R. and Sandberg, A. A. (1972). Estriol metabolism in sheep: Excretion of biliary and urinary conjugates. *Endocrinology*, **90**: 516–524.
272. Moodie, E. W. (1975). Mineral metabolism, pp. 63–99. (In reference 47.)
273. Moor, R. M. and Rowson, L. E. A. (1966). The corpus luteum of the sheep: Effect of the removal of embryos on luteal function. *J. Endocr.*, **34**: 497–502.
274. Morishima, H. O., Finster, M., Pedersen, H., Fukunaga, A., Ronfeld, R. A., Vassallo, H. G. and Covina, B. G. (1979). Pharmacokinetics of

Lidocaine in fetal and neonatal lambs and adult sheep. *Anesthesiology*, **50**: 431–436.

275. Morriss, F. H., Adcock, E. W., Paxson, C. L. and Greeley, W. J. (1979). Uterine uptake of amino acids throughout gestation in the unstressed ewe. *Am. J. Obstet. Gynecol.*, **135**: 601–608.
276. Mouw, D. R., Abrahams, S. F. and Mckenzie, J. S. (1974). Use of ventriculo-cisternal perfusion in conscious sheep. *Lab. anim. Sci.*, **24**: 505–509.
277. Munro, C. J., McNatty, K. P. and Renshaw, L. (1980). Circa-annual rhythms of prolactin secretion in ewes and the effect of pinealectomy. *J. Endocr.*, **84**: 83–89.
278. Murdoch, W. J., Lewis, G. S., Inskeep, E. K. and Tillson, S. A. (1978). The effect of a progesterone-releasing intrauterine contraceptive device on uterine secretion of prostaglandin $F_{2\alpha}$ and life-span of corpora lutea in the ewe. *Am. J. Obstet. Gynecol.*, **132**: 81–86.
279. Naughton, M. A., Meschia, G., Battaglia, F. C., Hellegers, A., Hagopian, H. and Barron, D. H. (1963). Hemoglobin characteristics and the oxygen affinity of the bloods of Dorset sheep. *Quart. J. exp. Physiol.*, **48**: 313–323.
280. Neill, A. R., Grime, D. W. and Dawson, R. M. C. (1979). Conversion of choline methyl groups through trimethylamine into methane in the rumen. *Biochem. J.*, **170**: 529–535.
281. Neill, A. R., Grime, D. W., Snoswell, A. M., Northrop, A. J., Lindsay, D. B. and Dawson, R. M. C. (1979). The low availability of dietary choline for the nutrition of the sheep. *Biochem. J.*, **180**: 559–565.
282. Nolan, J. V. (1975). Quantitative models of nitrogen metabolism in sheep, pp. 416–431. (In reference 246.)
283. Pagani, M., Mirski, I., Baig, H., Manders, W. T., Kerkhof, P. and Vatner, S. F. (1979). Effects of age on aortic pressure-diameter and elastic stiffness-stress relationships in unanesthetized sheep. *Circ. Res.*, **44**: 420–429.
284. Panaretto, B. A., Patterson, J. Y. F. and Hills, F. (1973). The relationship of the splanchnic, hepatic and renal clearance rates to the metabolic clearance rate of cortisol in conscious sheep. *J. Endocr.*, **56**: 285–294.
285. Panaretto, B. A. (1974). Relationship of visceral blood flow to cortisol metabolism in cold-stressed sheep. *J. Endocr.*, **60**: 235–245.
286. Pant, H. C., Hopkinson, C. R. N. and Fitzpatrick, R. J. (1977). Concentration of oestradiol, progesterone, luteinizing hormone and follicle stimulating hormone in the jugular venous plasma of ewes during the oestrous cycle. *J. Endocr.*, **73**: 247–255.
287. Paoni, N. F. and Castellino, F. J. (1979). A comparison of the urokinase and streptokinase activation properties of the native and lower molecular weight forms of sheep plasminogen. *J. biol. Chem.*, **254**: 2064–2070.
288. Parker, H. R., Dungworth, D. L. and Galligan, S. J. (1966). Renal hypertension in sheep. *Am. J. vet. Res.*, **27**: 430–443.
289. Patterson, J. Y. F., Bedford, C. A., Harrison, F. A. and Heap, R. B.

(1976). Progesterone and 20α-dihydroprogesterone in sheep: A model of their distribution and metabolism. *J. Reprod. Fertil.*, **46**: 313–322.

290. Patterson, J. Y. F. and Hills, F. (1967). The binding of cortisol by ovine plasma proteins. *J. Endocr.*, **37**: 261–268.

291. Pearson, L. D. and Brandon, M. R. (1976). Effect of fetal thymectomy on Ig G, Ig M and Ig A concentrations in sheep. *Am. J. vet. Res.*, **37**: 1139–1141.

292. Pearson, L. D., Doherty, P. C., Hapel, A. and Marshall, I. D. (1976). The responses of the popliteal lymph node of the sheep to Ross river and Kunjin viruses. *Aust. J. exp. Med. biol. Sci.*, **54**: 371–379.

293. Peart, J. N., Edwards, R. A. and Donaldson, E. (1972). The yield and composition of the milk Finnish Landrace × Blackface ewes: 1. Ewes and lambs maintained in doors. *J. agric. Sci.*, **79**: 303–313.

294. Penning, P. D. and Gibb, M. J. (1977). The use of corticosteroid to synchronize parturition in sheep. *Vet. Rec.*, **100**: 491–492.

295. Perrin, D. R. (1958). The chemical composition of the colostrum and milk of the ewe. *J. dairy Res.*, **25**: 70–74.

296. Phillips, R. W., House, W. A., Miller, R. A., Mott, J. L. and Sooby, D. L. (1969). Fatty acid, epinephrine, and glucagon hyperglycaemia in normal and depancreatized sheep. *Am. J. Physiol.*, **217**: 1265–1268.

297. Phillipson, A. T. (1970), editor. *Physiology of Digestion and Metabolism in the Ruminant*. Oriel Press; Newcastle-upon-Tyne.

298. Potter, B. J. (1958). Haemoglobinuria caused by propylene glycol in sheep. *Br. J. Pharmacol.*, **13**: 385–389.

299. Potter, B. J. (1963). The effect of saline water on kidney tubular function and electrolyte excretion in sheep. *Aust. J. agric. Res.*, **14**: 518–528.

300. Potter, B. J. (1972. The effect of prolonged salt intake on blood pressure in sheep. *Aust. J. exp. biol. med. Sci.*, **50**: 387–389.

301. Purser, D. B. and Buechler, M. (1966). Amino acid composition of rumen organisms. *J. Dairy Sci.*, **49**: 81–84.

302. Purvis, K., Illius, A. W. and Haynes, N. B. (1974). Plasma testosterone concentrations in the ram. *J. Endocr.*, **61**: 241–253.

303. Rawson, R. O. and Quick, K. P. (1972). Localization of intra-abdominal thermoreceptors in the ewe. *J. Physiol.*, **222**: 665–677.

304. Reid, R. L. (1951). Studies on the carbohydrate metabolism of sheep: 4. Hypoglycaemic signs and their relationship to blood glucose. *Aust. J. agric. Res.*, **2**: 146–157.

305. Renshaw, H. W. and Litteneker, N. M. (1978). Levels of total hemolytic complement activity in paired maternal newborn sheep sera. *Zentrabl. Vet. Med.* B, **25**: 689–694.

306. Rezakhani, A., Edjtehadi, M. and Szabuniewicz, M. (1977). Prevention of thiopental and thiopental/halothane cardiac sensitization to epinephrine in the sheep. *Can. J. comp. Med.*, **41**: 389–395.

307. Richard, P. (1967). *Atlas Stereotaxique du Cerveau de Brebis*. Institut national de la Recherche Agronomique; Paris.

308. Robertshaw, D. (1968). The pattern and control of sweating in the sheep and the goat. *J. Physiol.*, **198**: 531–539.
309. Robertson, H. A. (1977). Reproduction in the ewe and the goat, pp. 475–498. In *Reproduction in Domestic Animals* 3rd edition (edited by Cole, H. H. and Cupps, P. T.). Academic Press; New York.
310. Rollag, M. D., O'Callaghan, P. L. and Niswender, G. D. (1978). Serum melatonin concentrations during different stages of the annual reproductive cycle in ewes. *Biol. Reprod.*, **18**:279–285.
311. Rowson, L. E. A. (1971). Egg transfer in domestic animals. *Nature*, **233**: 379–381.
312. Ruckebusch, Y. and Bueno, L. (1977). Origin of migrating myoelectric complex in sheep. *Am. J. Physiol.*, **233**: E483–E487.
313. Ruckebusch, Y. and Thivend, P. (1980). *Digestive Physiology and Metabolism in Ruminants.* M.T.P. Press Ltd.; Lancaster, England.
314. Ryder, M. L. and Stephenson, S. K. (1968). See reference 35 in Chapter 1.
315. Samaja, M. and Gattinoni, L. (1978). Oxygen affinity in the blood of sheep. *Resp. Physiol.*, **34**: 385–392.
316. Sarelius, I. H. and Greenway, R. M. (1975). Rhythmic fluctuations in the urine composition of sheep: Separation of feed-dependent from other rhythms. *Pflugers Arch.*, **355**: 243–259.
317. Sawyer, M., Moe, J. and Osburn, B. I. (1978). Ontogeny of immunity and leucocytes in the ovine fetus and elevation of immunoglobulins related to congenital infection. *Am. J. vet. Res.*, **39**: 643–648.
318. Sawyer, M., Willadsen, C. H., Osburn, B. I. and McGuire, T. C. (1977). Passive transfer of colostral immunoglobulins from ewe to lamb and its influence on neonatal lamb mortality. *J. Am. vet. med. Assn.*, **171**: 1255–1259.
319. Scaramuzzi, R. J., Baird, D. T., Clarke, I. J., Martensz, N. D. and Van Look, P. F. A. (1980). Ovarian morphology and the concentration of steroids during the oestrous cycle of sheep actively immunized against androstenedione. *J. Reprod. Fertil.*, **58**: 27–35.
320. Schalm, O. W., Jain, N. C. and Carroll, E. J. (1975). *Veterinary Hematology*, 3rd edition. Lea and Febiger; Philadelphia.
321. Schanbacher, B. D. and Ford, J. J. (1976). Seasonal profiles of plasma luteinizing hormone, testosterone and estradiol in the ram. *Endocrinology*, **99**: 752–757.
322. Schaumloffel, E., Habermehl, A., Brock, N. and Schneider, B. (1973). Studies on the pharmokinetics of cyclophosphamide in sheep. *Arzneim.-Forsch.* (*Drug Res.*), **23**: 491–500.
323. Schmid, D. O., Cwik, S. and Forschner, J. (1975). Uber Lymphozytenantigene beim Schaf. *Zentrabl. vet. Med.*, **22B**: 386–392.
324. Schneider, N. and Yeary, R. A. (1975). Nitrite and nitrate pharmacokinetics in the dog, sheep and pony. *Am. J. vet. Res.*, **36**: 941–947.
325. Scoggins, B. A., Butkus, A., Coghlan, J. P., Denton, J. P., Fan, J. S.

K., Humphrey, T. J. and Whitworth, J. A. (1978). Adrenocorticotropic hormone-induced hypertension in sheep: A model for the study of the effect of steroid hormones on blood pressure. *Circ. Res.*, **43**: I–76–I–81. (But see Lun, S., Espner, E. A. and Hart, D. S. (1979). *Am. J. Physiol.*, **237**: E500–E503.)

326. Scott, D. (1969). Renal excretion of potassium and acid by sheep. *Quart. J. exp. Physiol.*, **54**: 412–422.
327. Scott, D. (1975). Changes in mineral, water and acid-base balance associated with feeding and diet, pp. 205–215. (In reference 246.)
328. Scott, T. W. and Cook, L. J. (1975). Effect of dietary fat on lipid metabolism in ruminants, pp. 510–523. (In reference 246.)
329. Setchell, B. P. and Waites, G. M. H. (1964). Blood flow and the uptake of glucose and oxygen in the testis and epididymis of the ram. *J. Physiol.*, **171**: 411–425.
330. Shafie, M. M. and Abdelghany, F. M. (1978). Structure of the respiratory system of sheep as related to heat tolerance. *Acta Anat.*, **100**: 441–460.
331. Shand, J. H. and Noble, R. C. (1974). δ/9 and δ/6-desaturase activities of the ovine placenta and their role in the supply of fatty acids to the foetus. *Biol. Neonate*, **36**: 298–304.
332. Shen, L. L., Barlow, G. H. and Holleman, W. H. (1978). Differential activities of heparins in human plasma and in sheep plasma: Effect on heparin molecular sizes and sources. *Thromb. Res.*, **13**: 671–679.
333. Shetty, S. N., Hines, J. A. and Edds, G. T. (1972). Effect of phenobarbital on bishydroxycoumarin plasma concentrations and hypoprothrombinemic responses in sheep. *Am. J. vet. Res.*, **33**: 825–834.
334. Simpson, H. and Edey, T. N. (1979). Changes in physical condition and ejaculate characteristics in paddock-mated rams. *Aust. vet. J.*, **55**: 225–228.
335. Simpson-Morgan, M. W. and Smeaton, T. C. (1972). The transfer of antibodies by neonates and adults, *Adv. vet. Sci.*, **16**: 355–386.
336. Slater, J. S. and Mellor, D. J. (1979). Concentrations of free amino acids in maternal and fetal plasma from conscious catheterized ewes during the last five weeks of pregnancy. *Res. vet. Sci.*, **26**: 296–301.
337. Smith, A., Calder, A. G., Lough, A. K. and Duncan, W. R. H. (1979). Identification of methyl-branched fatty acids from the triacylglycerols of subcutaneous adipose tissue of lambs. *Lipids*, **14**: 953–960.
338. Smith, F. G., Alexander, D. P., Buckle, R. M., Britton, H. G. and Nixon, D. A. (1972). Parathroid hormone in foetal and adult sheep: The effect of hypocalcaemia. *J. Endocr.*, **53**: 339–348.
339. Smith. J. B., Cunningham, A. J., Lafferty, K. J. and Morris, B. (1970). The role of the lymphatic system and lymphoid cells in the establishment of immunological memory. *Aust. J. exp. Biol. med. Sci.*, **48**: 57–70.
340. Smith, M. L., Lee, R., Sheppard, S. J. and Fariss, B. L. (1978). Reference ovine serum chemistry values. *Am. J. vet. Res.*, **39**: 321–322.

341. Smith, P. T. (1978). Electrocardiograms of 32 2-tooth Romney rams. *Res. vet. Sci.*, **24**: 283–286.
342. Smith, R. H. (1975). Nitrogen metabolism in the rumen and the composition and nutritive value of nitrogen compounds entering the duodenum, pp. 399–415. (In reference 246.)
343. Stacy, B. D., Brook, A. H. and Short, B. F. (1963). The relationship between suint and the sweat glands. *Aust. J. agric. Res.*, **14**: 340–348.
344. Stelmasiak, T. and Cumming, I. A. (1977). Two pools of pituitary L.H. An hypothesis explaining the control of the preovulatory surge of LH in the ewe. *Theriogenology*, **8**: 131.
345. Stewart-Tull, D. E. S. and Rowe, R. E. C. (1975). Procedures for large-scale antiserum production in sheep. *J. Immunological Methods*, **8**: 37–46.
346. Sutherland, R. L. and Simpson-Morgan, M. W. (1975). The thyroxine-binding properties of serum proteins. A competitive binding technique employing Sephadex G25. *J. Endocr.*, **65**: 319–332.
347. Symons, L. E. A. and Jones, W. O. (1966). The distribution of dipeptidase activity in the small intestine of the sheep (*Ovis aries*), *Comp. Biochem. Physiol.*, **18**: 71–82.
348. Tao, R. C. Asplund, J. M. and Kappel, L. C. (1974). Response of nitrogen metabolism, plasma amino acids and insulin levels to various levels of methionine infusion in sheep. *J. Nutr.*, **104**: 1646–1656.
349. Tarttelin, M. F. (1976). The ventromedial nucleus of the hypothalamus of sheep (*Ovis aries*) and the effects on food and water intake following its electrolytic destruction. *Acta Anat.*, **94**: 248–261.
350. Taylor, R. B. (1962). Pancreatic secretion in the sheep. *Res. vet. Sci.*, **3**: 63–77.
351. Theil, E. C. and Calvert, K. T. (1978). The effect of copper excess on iron metabolism in sheep. *Biochem. J.*, **170**: 137–143.
352. Thompson, R. H. and Todd, J. R. (1976). Role of pentose-phosphate pathway in haemolytic crisis of chronic copper toxicity of sheep. *Res. vet. Sci.*, **20**: 257–260.
353. Tomas, F. M. (1974) The influence of parathyroid hormone on the secretion of phosphate by the parotid salivary gland of sheep. *Quart. J. exp. Physiol.*, **59**: 269–281.
354. Tomas, F. M. (1975). Renal response to intravenous phosphate infusion in the sheep. *Aust. J. biol. Sci.*, **28**: 511–520.
355. Tomas, F. M. and Somers, M. (1974). Phosphorus homeostasis in sheep: 1. Effects of ligation of parotid salivary ducts. *Aust. J. agric. Res.*, **25**: 475–483.
356. Toutain, P.-L., Toutain, C., Webster, A. J. F. and McDonald, J. D. (1977). Sleep and activity, age and fatness, and the energy expenditure of confined sheep. *Br. J. Nutr.*, **38**: 445–454.
357. Traber, D. L. *et al.* (1979). See reference 43 in Chapter 1.
358. Trautman, A. and Fiebiger, J. (1952). *Fundamentals of the Histology of*

Domestic Animals. Translated by Habel, R. E. and Biberstein, E. L. Comstock Publishing Associates; Ithaca, New York.
359. Trenkle, A. (1976). Estimates of the kinetic parameters of growth hormone metabolism in fed and fasted calves and sheep. *J. anim. Sci.*, **43**: 1035–1043.
360. Tsang, C. P. W. (1978). Plasma levels of estrone sulfate, free estrogens and progesterone in the pregnant ewe throughout gestation. *Theriogenology*, **10**: 97–110.
361. Tsang, C. P. W. and Hackett, A. J. (1979). Metabolism of estrone sulphate in the pregnant sheep. *Theriogenology*, **11**: 429–439.
362. Tsang, C. P. W. and Hackett, A. J. (1979). Metabolism of progesterone in the pregnant sheep near term: Identification of 3β-hydroxy-5α-pregnan-20-one 5-sulfate as a major metabolite. *Steroids*, **33**: 577–588.
363. Tucker, A., McMurtry, I. F., Reeves, J. T., Alexander, A. F., Will, D. H. and Grover, R. F. (1975). Lung vascular smooth muscle as a determinant of pulmonary hypertension at high altitude. *Am. J. Physiol.*, **228**: 762–767.
364. Tucker, E. M. (1971). Genetic variation in the sheep red blood cell. *Biol. Rev.*, **46**: 341–386.
365. Tucker, E. M. (1974). A shortened life span of sheep red cells with a glutathione deficiency. *Res. vet. Sci.*, **16**: 19–22.
366. Tucker, E. M. (1975). Genetic markers in the plasma and red blood cells, pp. 123–153. (In reference 47.)
367. Tucker, E. M. (1976). Some physiological aspects of genetic variation in the blood of sheep. *Anim. Blood Grps. biochem. Genet.*, **7**: 207–215.
368. Tucker, E. M. and Crowley, C. (1978). NADH diaphorase as a genetic marker for sheep red cells. *Anim. Blood Grps. biochem. Genetics*, **9**: 161–167.
369. Tucker, E. M., Kilgour, L. and Young, J. D. (1976). The genetic control of red cell glutathione deficiencies in Finnish Landrace and the Tasmanian Merino sheep and in crosses between these breeds. *J. agric. Sci.*, **87**: 315–323.
370. Tucker, E. M., Moor, R. M. and Rowson L. E. A. (1974). Tetraparental sheep chimeras induced by blastomere transplantation: Changes in blood type with age. *Immunology*, **26**: 613–621.
371. Tucker, E. M. and Young, J. D. (1976). Genetic variation in the purine nucleoside phosphorylase activity of sheep red cells. *Anim. Blood Grps. biochem. Genetics*, **7**: 109–117.
372. Turner, A. W. and Hodgetts, V. E. (1959). The dynamic red cell storage function of the spleeen in sheep: 1. Relationship to fluctuations of jugular haematocrit. *Aust. J. exp. Biol. med. Sci.*, **37**: 399–420.
373. Turner, H. N. and Young, S. S. Y. (1969). *Quantitative Genetics in Sheep Breeding*. Macmillan; Australia. (See also Turner, H. N. in *Anim. Breeding Abstr.*, **37**: 545–563 (1969) and **40**: 621–634 (1972)).
374. Ulyatt, M. J., Dellow, D. W., Reid, C. S. W. and Bauchop, T. (1975). Structure and function of the large intestine of ruminants, pp. 119–133. (In reference 246.)
375. Ungerer, T., Orr, J. A., Bisgard, G. E. and Will, J. A. (1976).

Cardiopulmonary effects of mechanical distension of the rumen in nonanesthetized sheep. *Am. J. vet. Res.*, **37**: 807–810.
376. Vegad, J. L. (1979). The acute inflammatory response in the sheep. *Vet. Bull.*, **49**: 555–561.
377. Vegad, J. L. and Lancaster, M. C. (1972). Cutaneous antigen-antibody reactions in the sheep. *N.Z. vet. J.*, **20**: 103–108.
378. Voglmayr, J. K., Roberson, C. and Musto, N. A. (1980). Comparison of androgen levels in ram rete testis fluid, testicular lymph and spermatic venus blood plasma: Evidence for a regulatory mechanism in the seminiferous tubules. *Biol. Reprod.*, **23**: 29–39.
379. Wade, L. (1973). Splenic sequestration of young erythrocytes in sheep. *Am. J. Physiol.*, **224**: 265–267.
380. Wade, L. and Sasser, L. B. (1970). Body water, plasma and erythrocyte volume in sheep. *Amer. J. vet. Res.*, **31**: 1375–1378.
381. Waites, G. M. H. (1957). The course of the efferent cardiac nerves of the sheep. *J. Physiol.*, **139**: 417–433. (See also McKibben, J. S. and Getty, R. (1969). *Acta Anat.*, **74**: 228–242.)
382. Waites, G. M. H. (1961). Polypnoea evoked by heating the scrotum of the ram. *Nature*, **190**: 172–173.
383. Waites, G. M. H. and Moule, G. R. (1960). Blood pressure in the internal spermatic artery of the ram. *J. Reprod. Fertil.*, **1**: 223–229.
384. Walsh, S. W., Yutrzenka, G. J. and Davis, J. S. (1979). Local steroid concentrating mechanism in the reproductive vasculature of the ewe. *Biol. Reprod.*, **20**: 1167–1171.
385. Walton, J. S., Evins, J. D., Fitzgerald, B. P. and Cunningham, F. J. (1980). Abrupt decrease in day length and short-term changes in the plasma concentrations of FSH, L.H. and prolactin in anoestrous ewes. *J. Reprod. Fertil.*, **59**: 163–171.
386. Wanner, A., Mezey, R. J., Reinhart, M. E. and Eyre, P. (1979). Antigen-induced bronchospasm in conscious sheep. *J. appl. Physiol. Respirat: Environ. Exercise Physiol.*, **47**: 917–922.
387. Warner, A. C. I. and Stacy, B. D. (1972). Water, sodium and potassium movements across the rumen wall of sheep. *Quart. J. exp. Physiol.*, **57**: 103–119.
388. Warner, A. C. I. and Stacy, B. D. (1977). Influence of ruminal and plasma osmotic pressure on salivary secretion in sheep. *Quart. J. exp. Physiol.*, **62**: 133–142.
389. Weber, K. M., Boston, R. C. and Leaver, D. D. (1980). A kinetic model of copper metabolism in sheep. *Aust. J. agric. Res.*, **31**: 773–790.
390. Webster, A. J. F., Smith, J. S., and Brockway, J. M. (1972). Effects of isolation, confinement and competition for feed on the energy exchanges of growing lambs. *Anim. Prod.*, **15**: 189–201.
391. Webster, M. E. D. Department of Physiology, University of New England, Australia. Personal communication.
392. Webster, M. E. D. and Johnson, K. G. (1968). Some aspects of body temperature regulation in sheep. *J. agric Sci.*, **71**: 61–66.

393. Welento, J., Szteyn, S. and Millart, Z. (1969). Observations on the stereotaxic configuration of the hypothalamus nuclei in the sheep. *Anat. Anz.*, **124**: 1.

394. Wenham, G. and Wyburn, R. S. (1980). A radiological investigation of the effects of cannulation on intestinal motility and digesta flow in sheep. *J. agric. Sci.*, **95**: 539–546.

395. Whitworth, J. A., Coghlan, J. P., Denton, D. A., Hardy, K. J. and Scoggins, B. A. (1979). Effect of sodium loading and ACTH on blood pressure of sheep with reduced renal mass. *Cardiovasc. Res.*, **13**: 9–15.

396. Whyman, D., Johnson, D. L., Knight, T. W. and Moore, R. W. (1979). Intervals between multiple ovulations in PMSG-treated and untreated ewes and the relationship between ovulation and oestrus. *J. Reprod. Fertil.*, **55**: 481–488.

397. Wilkes, P. R., Munro, I. B. and Wijeratne, W. V. S. (1978). Studies on a sheep freemartin. *Vet. Rec.*, **102**-140–142.

398. Willadsen, S. M. (1979). A method for culture of micromanipulated sheep embryos and its use to produce monozygotic twins. *Nature*, **277**: 298–300.

399. Willadsen, S. M. (1980). The viability of early cleavage stages containing half the normal number of blastomeres in the sheep. *J. Reprod. Fertil.*, **59**: 357–362.

400. Willadsen, S. M., Polge, C., Rowson, L. E. A. and Moor, R. M. (1976). Deep freezing of sheep embryos. *J. Reprod. Fertil.*, **46**: 151–154.

401. Wilson, K. C., DiStefano, J. J., Fisher, D. A. and Sack, J. (1977). System analysis and estimation of key parameters of thyroid hormone metabolism in sheep. *Ann. Biomed. Eng.*, **5**: 70–84.

402. Wilson, P. R. and Tarttelin, M. F. (1978). Studies of sexual differentiation of sheep: 1. Foetal and maternal modifications and post-natal plasma LH and testosterone content following androgenization early in gestation. *Acta Endocr.*, **89**: 182–189. (See also **89**: 190–195).

403. Winter, H. (1964). The myelogram of normal sheep. *J. comp. Path.*, **74**: 457–469.

404. Wodzicka-Tomaswezska, M., Hecker, J. F. and Bray, A. R. (1974). Effects of day of insertion of intrauterine devices on luteal function in ewes. *Biol. Reprod.*, **11**: 79–84.

405. Woods, J. R., Dandavino, A., Murayama, K., Brinkman, C. R. and Assali, N. S. (1977). Autonomic control of cardiovascular functions during neonatal development and in adult sheep. *Circ. Res.*, **40**: 401–407.

406. Woods, J. R., Dandavino, A., Nuwayhid, B., Brinkman, C. R. and Assali, N. S. (1978). Cardiovascular reactivity of neonatal and adult sheep to autonomic stimuli during adrenergic depletion. *Biol. Neonat.*, **34**: 112–120.

407. Wright, A. A. (1962). Metabolism of Oestradiol-17 β by an ovariectomized ewe. *J. End.*, **24**: 291–297.

408. Wright, P. C ., Young, J. D., Mangan, J. L. and Tucker, E. M. (1977). An inherited arginase deficiency in sheep erythrocytes. *J. agric. Sci.*, **88**: 765–767.

409. Yamamoto, Y., Peric-Golia, L., Osawa, Y., Kirdani, R. Y. and Sandberg, A. A. (1978). Androgen metabolism in sheep. *Steroids*, **32**: 373–388.
410. Yesberg, N. E., Henderson, M. and Budtz-Olsen, O. E. (1978). The effects of two analogues of arginine-vasopressin (ornithine-vasopressin and desamino-d-arginine vasopressin) on kidney function in sheep. *Quart. J. exp. Physiol.*, **63**: 179–188.
411. Young, B. A. (1966). Energy expenditure and respiratory activity of sheep during feeding. *Aust. J. agric. Res.*, **17**: 355–362.
412. Young, B. A., Bligh, J. and Louw, G. (1976). Effect of thermal tachypnoea and of its mechanical or pharmacological inhibition on hypothalmic temperature in the sheep. *J. thermal Biol.*, **1**: 195–198.
413. Young, J. E., Younger, R. L., Radeleff, R. D., Hunt, L. M. and McLaren, J. K. (1965). Some observations on certain serum enzymes of sheep. *Am. J. vet. Res.*, **26**: 641–644.
414. Zanjani, E. D., McGlave, P. B., Bhakthavathsalan, A. and Stamatoyannopoulos, G. (1979). Sheep fetal haematopoietic cells produce adult haemoglobin when transplanted in the adult animal. *Nature*, **280**: 495–496.
415. Zehr, J. E., Johnson, J. A. and Moore, W. W. (1969). Left atrial pressure, plasma osmolality and ADH levels in the unanaesthetized ewe. *Am. J. Physiol.*, **217**: 1672–1680.
416. Zimmerman, M. B., Blaine, E. H. and Stricker, E. M. (1981). Water intake in hypovolemic sheep: Effects of crushing the left atrial appendage. *Science*, **211**: 489–491.
417. Zimmerman, H. J., Schwartz, M. A., Boley, L. E. and West, M. (1965). Comparative serum enzymology. *J. lab. clin. Med.*, **66**: 961–972.

Chapter 4

Behaviour

4.1 Behaviour in the Field

Before considering the behaviour of sheep in the laboratory, it is of value to outline their behaviour under field conditions as this information will provide understanding of behaviour in the laboratory. A summary of the behavioural characteristics of sheep is given in Table 36. Much of the information given in this section is from Hafez *et al.*[18] (see also [2]).

Of primary importance is the fact that the sheep has a highly developed flocking or herding instinct. This and its pronounced docility were probably the most important factors in its domestication. Fossil records show that the sheep and the goat were domesticated early in man's history.

The flocking instinct is well seen in the grazing patterns of sheep where they usually move over an area in groups rather than as individuals. British bred sheep tend to be more independent as they are accustomed to roaming over bracken covered hills looking for the occasional blade of grass. Other breeds and especially the Merino graze in sizeable flocks. There are no "leaders" in the flock which initiate grazing or other forms of flock behaviour.

Grazing can occur at any time of the day or night but is usually most intense early in the morning and late in the afternoon until dusk. Typically sheep will have between four and seven grazing periods each day of which some will be at night. Night-time grazing occurs less often in winter. The total time spent grazing is on average 10 h per day.

The sheep has a cleft upper lip which is less prehensile than the

Table 36 Behaviour patterns of sheep (from [18]).

Behaviours	Characteristic pattern
Ingestive	Grazing, browsing, ruminating, licking salt, drinking. Suckling-nudging udder with nose, sucking, wiggling tail.
Shelter-seeking	Moving under trees, into barns. Huddling together to keep off flies. Crowding together in extreme cold weather. Pawing ground and lying down.
Investigatory	Raising head, directing eyes, ears, and nose towards disturbance. Nosing an object or another sheep.
Allelomimetic	Walking, running, grazing, and bedding down together. Following one another. Bouncing stiff-legged past an obstacle together.
Agonistic	Pawing, butting, shoving with shoulders. Running together and butting. Brunching and running. Lying immobile.
Eliminative	Urination posture: squatting of females, arching back, bending legs (male lambs). Defaecation: wiggling tails.
Care-giving (Epimeletic) (females only)	Licking and nibbling placental membranes and young. Arching back to permit nursing, nosing lamb at base of tail. Circling new-born lamb. Baaing when separated from lamb.
Care-soliciting	Baaing (distress vocalization by lamb when separated, hungry, hurt, or caught). Baaing of adults when separated from flock.
Sexual (male)	Courtship: following female, pawing female, hoarse baa or grumble, nosing genital region of female, sniffing female urine, extending neck with uncurled lip, running tongue in and out, rubbing against side of female, biting wool of female, herding or pushing female away from other males. Copulation: wiggling tail, mounting, thrusting movements of hind quarters.
Sexual (female)	Courtship: rubbing neck and body against male, mounting male (rare). Copulation: standing still to receive male.

(*Table continued overleaf*)

Table 36 Behaviour patterns of sheep (from [18]).—continued

Behaviours	Characteristic pattern
Play	Sexual: mounting by either sex. Agonistic: playful butting. Allelomimetic: running together, "gambolling" (bouncing stiff-legged and turning in air) together. "Game playing"—jumping off and on rock together.

non-cleft upper lip of the cow. Sheep tend to graze close to the ground and do not like long grass. In eating, they hold the grass between the incisor teeth on the upper dental pad and jerk the head slightly forwards and upwards so as to break off the blades of grass. They are selective in their grazing and have definite preferences for certain grass species.

Another 8 to 10 h per day is spent in ruminating. A sheep will have between 8 and 15 rumination periods each day, each of which will last for between 2 min and 2 h. Typical values for rumination are 90 chews/min, 78 chews/bolus and 500 boluses/day[18] although this will depend on the food intake and the state of the roughage component.[36] Stimuli for starting and for ending periods of rumination are unknown. During rumination, the sheep normally sitting with its head held up will regurgitate a bolus of food from the rumen and start chewing with strong regular side to side jaw movements. Chewing is almost always in one direction. Only the molar and premolar teeth are used and one bolus will be held on one side of the mouth or the other during the whole time that it is chewed. Chewing suddenly stops, the bolus is swallowed and a few seconds later another bolus is regurgitated.

Urination is performed some 9 to 13 times each day and defaecation 6 to 8 times per day. Three to 6 litres of water are drunk daily. However experiments have shown that grazing sheep other than lactating ewes can live in temperate climates for several years without access to drinking water.[27] Under such circumstances, they tend to reduce the grazing time in the heat of the day and increase grazing at night and in the early morning when dew is on the grass.

Most breeds of sheep will walk 8 to 16 km each day, even when kept in a small field, but the British hill breeds and the Merino tend to walk longer distances. When not grazing, Merino sheep tend to camp at specific camp sites which are often on hills or against fences where they can meet with sheep from adjoining fields. In hot weather, camps are under trees for shade. Sheep camp sites are obvious because of growth from weed seeds carried there in faeces and caught in wool. Also defaecation and urination at the sheep camp sites increase the fertility of the soil so that the vegetation has a darker green. As a result of these factors, camp sites tend to grow weeds, the surrounds clover and elsewhere grass. Hill breeds in their normal habitat may not camp together but individual sheep will shelter from the wind behind large rocks and tussocks of grass.

When two flocks of sheep are introduced into a paddock and are of the same sex, age and breed, initially they do not mix but take about 3 weeks to integrate. When they are of the same sex and age but of different breeds, they may not integrate even after a period of 2 years in the same field. This indicates that sheep recognize animals of the same or similar breeds and also can distinguish them from sheep of different breeds.[4] Most sheep are raised in flocks of one breed and it seems that this does not apply to flocks which are initially of mixed breeds.[22] In a similar light, Clun Forest rams prefer to mate with Clun Forest ewes and Merino rams with Merino ewes.[25]

Reproductive behaviour is important in sheep production. Mating of rams with ewes in oestrus is most frequent in the early morning or in the evening. Rams when searching for ewes in oestrus walk around the flock nosing the perineums and nudging ewes. Vision is an important sense for a ram to detect a ewe in oestrus but smell is also important.[14] If a ram suspects that a ewe might be in oestrus, he will usually show the Flehmen phenomenon (Fig. 13). The ram may also paw at the ewe with one of his forelegs while standing alongside the ewe. If the ewe shows a negative response (see below), the ram will usually move to another ewe.

In the absence of a ram, it is difficult for a person to detect that a ewe is in oestrus. In the presence of a ram, such detection is easy. The ewe will often tend to seek out the ram and, if in a field next to the one containing a ram, will usually be beside or near the fence separating the two fields and in the vicinity of the ram. If the ram is tethered in a field, most but not all ewes will mate with him, illustrating that the ewe's behavioural role is positive.[26]

In response to a ram smelling the perineum, a ewe will give one of two signs. She may stand, turn her head to one side to look at the ram

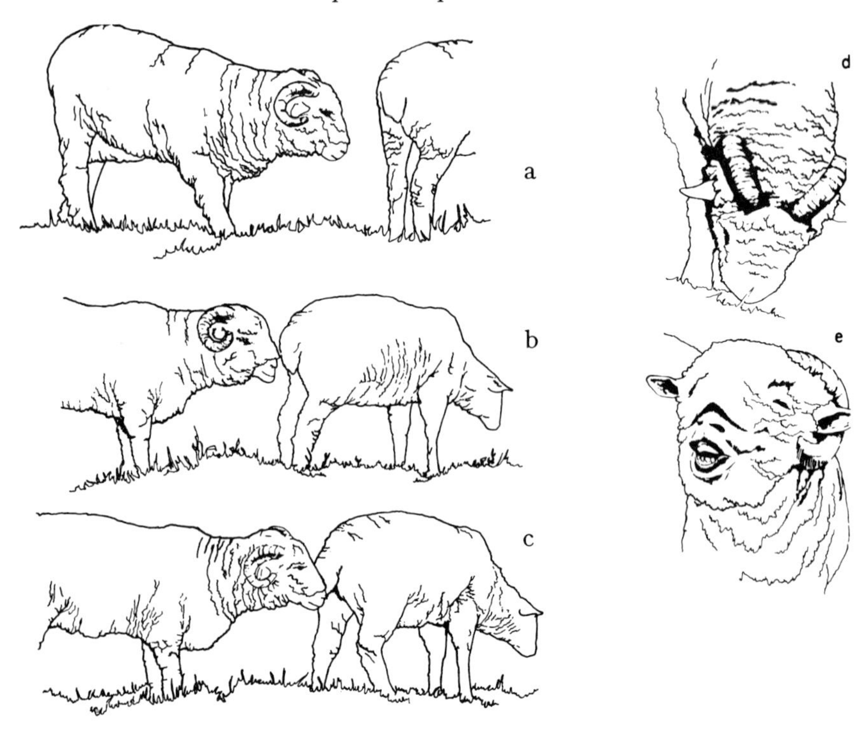

Fig. 13 *Courtship in sheep*. Ram approaches ewe (a) and noses her perineum (b); ewe urinates (c); and ram noses urine on ground (d). Ram showing Flehman response (e).
[From Banks, E. M. (1964) (*Behaviour* **23**: 249–279) by permission of E. J. Brill, Leiden.]

and wiggle her tail. This is a positive sign that she is interested. Alternatively she may give a negative sign by either squatting slightly, passing a small amount of urine and then walking away or just walking away. At the start or the end of an oestrous period, the ewe's reaction to the ram can be variable.

After a short "foreplay" which may include rubbing his head along the side of the ewe, licking and biting the ewe's wool and making noises, the ram will mount and thrust several times. The ewe may hold up her tail to aid his penetration. A ram might have several unsuccessful mounts but once he achieves intromission he will make a few small thrusts and then a characteristic deep pelvic thrust with a backwards movement of the head. Then he will demount and stand with his head lowered for a short while before walking away. It will be a few minutes before he is interested in serving again.

Introduction of rams to a flock of ewes will cause many to come into oestrus and to ovulate within a few days. A pheronome is involved as extracts of wool or wax collected from around the eyes or bare skin of the flanks of rams will also stimulate cycling in ewes.[24] Introduction of a ram to ewes at or near the start of the normal breeding season can also hasten the initiation of oestrous cycling in ewes.

Rams vary in their sexual capacity and libido[29,52] and an occasional young ram may show no interest in ewes. He is likely to have been reared in an all male group and may have preferences for males as sexual partners rather than for ewes.[53] A measure of libido is the number of ewes that a ram will serve in a day. Libido tends to have a significant hereditary component but is also influenced by exposure to androgens before puberty. Testosterone implants given to the ram lamb during the second month after birth will decrease the proportion of rams exhibiting low frequency of service as adults.[30]

An occasional ram will have low libido or prowess. This usually will not be noticed if he is one of a number of rams used in a flock. Such a ram could cause problems if he is used alone with a small number of ewes in a laboratory flock.

Apart from swelling of the udder and relaxation of the vulva, ewes given few signs of impending parturition until just before the event. It has been suggested that there is a drop in temperature of the ewe of about 0·3°C or greater on the morning of parturition.[50] Often ewes will separate from the flock a short time before birth. They may appear unsettled or agitated and often paw the ground with a forefoot as if trying to clear a site for lambing. This however may just be the natural tendency for sheep to paw the ground when disturbed or uncomfortable. At this stage, they can steal a lamb from a ewe which has just delivered one lamb (see imprinting below) and may be giving birth to another lamb.

Parturition is rapid, taking for one lamb usually about 1½ h between onset of labour and birth. It can be as short as 20 min or may take much longer, especially if the lamb is large in proportion to the size of the pelvis. Some breeds with large heads (in particular the Dorset Horn) have a high incidence of dystocia. For parturition, the ewe usually sits on the ground and may roll partly on to one side. When there is a multiple pregnancy, the interval between successive lambs is about half an hour and this period allows the ewe to clean up the previous lamb and allows it to suckle. Once a lamb is born, the ewe soon gets to her feet and licks the lamb vigorously starting from the head. The ewe will help the lamb in its attempts to rise to its feet. Once the lamb is on its feet, it moves to the ewe and passes backwards along her underside trying to suck from protruding pieces of wool until it finds a teat from which it starts sucking. Many ewes will eat the afterbirth which is normally passed a few minutes after the birth of the lamb.

Imprinting, the formation of a bond between the ewe and lamb, occurs soon after birth. The lamb and ewe learn to pick out each other's sounds, smells and colours and these factors allow each to find the other within the multitude of ewes and lambs in a flock. Sound is the most important factor over distances of 10 m or more. A ewe can recognize auditory signals from a recording made several days previously.[38] At a distance of less than 1 or 2 ft, smell is the predominant factor for recognition of the lamb.[32]

The critical time for imprinting for the ewe is within the first few hours although it is longer for the lamb. Licking for a period of 20 to 30 min appears to provide the bond of the ewe to the lamb. A lamb removed at birth will usually reunite with its mother if returned within about 8 h.[45] It may be necessary in an experiment to remove a lamb for a short period immediately after birth or to foster an orphan lamb onto a ewe that has lost her own lamb. Neathery[34] suggests use of tranquillizers for the ewe to gain acceptance of the lamb. These probably provide a period of a few hours when the ewe will accept any lamb during which time the lamb will bond to the ewe and learn to suckle properly. When the effect of the tranquillizers has worn off, persistence by the lamb will gradually gain acceptance by the ewe.*

Careful studies have shown that stealing of lambs by immediately preparturient ewes may reach an incidence of up to 25%.[1,47] Stealing is most common when ewes are lambed in close confinement in large numbers and when lambing has been synchronized (p. 147). Stealing should be kept in mind in breeding experiments with sheep.

Lambs suckle up to 70 times per day during the first week, each time for 1 to 3 min. Later, the time is reduced to 30 s each time. This is insufficient for the milk let-down reflex to occur and so the lamb must take only the milk present in the larger ducts of the mammary gland.

* Alternatively, a lamb "adapter"[48] which allows a lamb access to the udder of a ewe for a period of 24 h may be used.

Lambs will eat small amounts of grass and will ruminate for short periods a day or two after birth. As the lamb grows older, the time spent eating and spent ruminating increases. The ewe will wean the lamb after 4 to 6 months.

In all groups of animals, social hierarchies, otherwise known as "peck orders", are established. In a social hierarchy, animals tend to become ordered in their dominance over other animals in the group so that there is one most dominant animal and other animals with lesser degrees of dominance including one at the bottom of the hierarchy which is not dominant to any other in the group. In ewes and wethers, social hierarchy is not very evident as it would conflict with the well-developed flocking interest. Dominance is well developed in rams and when two rams first meet, they will test their strength by charging at each other head down with a resultant collision that would seem to cause great damage but surprisingly rarely does. Rams will repeat these violent encounters several times until one decides that he has had enough and that he accepts the dominance of the other ram. Because rams can charge, caution should be exercised when approaching rams from the front in a race or small yard as one might just charge to escape. It is safe to approach a ram from the rear or in a field where they have sufficient space to turn around and move away.

Dominance can be detected in ewes and wethers by careful observation, especially if they are under stress such as being kept in crowded conditions. In a yard, the most dominant sheep of a group is likely to be furthest from people and the least dominant closest.[10] A situation where dominance and subordination are clearly expressed is when there is competition in a group pen for space at a food trough.[3]

Poddy lambs (lambs reared by hand) are likely to show differences in behaviour from lambs reared naturally. These differences are likely to be marked if such lambs have been reared with little contact with other sheep.

4.2 Behaviour in the Laboratory

Less is known about behaviour of sheep under laboratory conditions. In cold climates, sheep are often held under cover in close confinement for the winter months. They adjust well to these conditions and also appear to adjust well to conditions of close confinement in the laboratory.

Ruckebusch[41] studied the state of sleep in laboratory sheep. They were awake for about 16 h/day (more during day time than at night) and "drowsy" for abour 4 h (more at night than during the day). Sleep (4 h) took two forms. Slow wave sleep occurred mainly at night and paradoxical sleep only at night. They stood for 70% of the 24 h and slept for some of the time while standing. Standing occurred more in the day time than at night. Rumination took place while awake and while drowsy. There may be a relation between rumination and sleeping as Morag[32] observed a marked increase in sleeping when sheep were fed on a very finely ground roughage diet which suppressed rumination. He found it necessary to prod sheep to waken them.

An ethogram of the behaviour of sheep in a laboratory animal house is shown in Fig. 14.

It is common for zoo animals to show stereotyped behaviour. This is the repeated performance of a movement or a series of movements which have no discernible purpose or advantage to the animal in that situation and which are not normally performed in a normal habitat. In a study of sheep recently brought into an animal house compared with sheep that had been there for several months, Done[9] observed that patterns of stereotype behaviour not seen in newly introduced sheep were obvious in sheep that had been confined for about 6 months and very prominent in sheep confined for up to 2 years. More time (up to 3 h in a 10 h period) was spent in stereotype movements on weekdays when there was human activity in the animal house than on Sundays when human activity was minimal. Set patterns were often performed when sheep became "excited", such as before feeding and when strangers came close. But they were also performed when no humans were present and there was no disturbance. In contrast to these sheep which had been in the animal house for long periods, newly introduced sheep tended to be withdrawn and not to react to activities of workmen.

Although dominance patterns are not well developed in sheep, they may cause an occasional problem in sheep kept in group pens in an animal house for a long period. A dominant sheep may "pick" on a subordinate sheep and repeatedly charge it. Alternately it may start chewing the subordinate animal's wool. Very occasionally, it may be necessary to remove the dominant or subordinate animal to another pen.

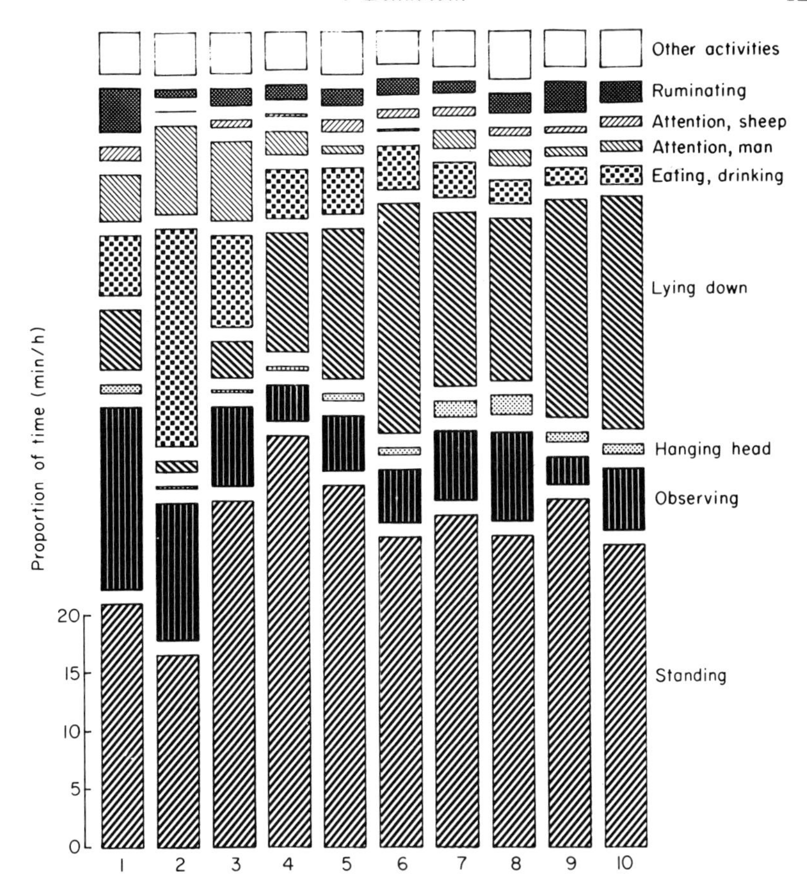

Fig. 14 *Time spent at various activities by sheep in an animal house.* [From Done.[9]]

A question of relevance to this book is the time that it takes for a sheep to settle into a laboratory routine. Most research workers allow a period of 1 to 2 weeks before performing experiments on conscious sheep. A sheep will probably be accustomed to the laboratory routine by this time but will not be completely "tamed" as it will still show a transitory rise in adrenal corticosteroids on being handled.[37] However after a period of 1 to 2 weeks, the basal adrenal corticosteroid level which will have risen on introduction to the animal house will have returned to normal.[37]

An untamed sheep when introduced to the animal house may show "wild" behaviour. When approached, it will remain alert at the back of the pen and may stamp a forefoot or paw the floor. It may also circle rapidly and may even attempt to jump out of the pen. Its reactions on being approached will be less if the flock from which it has been taken has had much contact with humans in the field. With a good training regime, this behaviour pattern can be changed to one where the sheep shows no fear on the approach of and handling by a person with whom it is familiar. It may even come up to the experimenter to have a blood sample taken if at the same time it is offered a little food in the hand and there will be no rise in the blood adrenal corticosteroid level.[37] The period for this training may take 3 to 6 weeks depending on the breed and previous handling. Initially some tasty food such as lucerne (alfalfa) chaff or dried molasses should be offered every time that the handler walks near the sheep. If the sheep will not readily sniff the hand which offers the food, the food should be placed in the food container and the handler should then step back until the sheep realizes that the outstretched hand is associated with food. The handler can then start stroking the sheep on the head and neck for a minute or two before giving the handful of food. Some 5 to 10 min per day should be sufficient. A further short period of training (3 or 4 days) may be necessary to settle the sheep into the actual experimental routine in the laboratory.

Although some investigators may use only one well-trained sheep at a time in the laboratory (space may be a factor), it is preferable to have two sheep placed so that they are face to face or at least can see each other. In the absence of a second sheep, a large mirror hung in front of the sheep crate may be sufficient to provide reassurance to the sheep that it is not alone.

In one institute, it was the custom for the animal house staff to wear grey or kahki protective dress and the research staff to wear white. The sheep tended to react to the white colour until the animal house staff changed the colour of their protective dress to white.

4.3 Reactions of Sheep to Stress

External cues may provide a good indication of when an animal is subject to acute stress but not when the stress is mild or subtle. There are several reports in the literature of changes in physiological parameters in response to factors associated with experimentation.

Some of these will be described to give insight into the reactions of sheep to various stimuli so that experiments may be planned to minimize these reactions.

Obvious reactions to stress are an increased heart rate and a contraction of the spleen which causes an increased haematocrit (p. 34). Changes in levels of hormones such as cortisol provide good indications of stress, as suggested above. Transferring sheep to small cages caused increased mean levels and erratic peaks in blood cortisol which subsided after 6 days.[19] Cortisol levels are higher in ewes involved in longer or more difficult parturition.[6] When a sheep is subjected to an acute stressful situation such as shearing or bringing a barking dog near, some 30 to 40 min or more elapse before the maximal cortisol levels are reached in blood.[12,23] There may even be an initial drop in cortisol secretion rate.[12] The secretion rate of cortisol may take 6 h to return to normal.[12] Prolactin levels are also markedly elevated by such procedures as restraint and venepuncture.[46]

Adrenaline is another stress-related hormone and it has been established that adrenaline given intravenously causes a dose related increase in plasma cortisol levels[31] by stimulation of the release of ACTH from the anterior pituitary gland. In this experiment, the cortisol response to adrenaline was lower in sheep that were well accustomed to the laboratory conditions.

Adrenaline and/or cortisol have been associated with some other changes such as an increase in blood glucose concentration of up to 17% in sheep when jugular catheters were inserted.[49] Catheterization by a stranger can cause up to a five fold increase in blood lactate and changes in other metabolites.[17] Even in a sheep accustomed to sampling from a catheterized jugular vein, a small drop in temperature will be seen in a "sympathectomized" ear, indicative of release of adrenaline.[42] An infusion of adrenaline will cause a marked drop in the plasma magnesium level and it has been suggested that sudden stress resulting in adrenaline secretion may be the trigger to the clinical condition of hypomagnesaemia in sheep on a magnesium deficient diet.[39] Increased cortisol causes increased excretion of copper in bile.[7] At least one of these hormones released in apprehension could be the cause of increased blood lactate seen prior to one experiment.[40]

A further subtle indicator of stress is lymphatic function as illustrated in an experiment in which "acute painful stimuli" were administered to sheep with lymph duct fistulae.[43] These stimuli were the insertion of an 18 gauge hypodermic needle subcutaneously or the injection of 5 ml of a streptomycin-penicillin mixture intramuscularly and are no more painful than some common laboratory procedures. They caused increases in both the lymph flow rate and cell content within 15 min and the effects had subsided by 30 min. These effects were not reproduced by intravenous injections of adrenaline.

Stresses such as close proximity to a barking dog and cannulation of the jugular vein also cause release of thyroid hormones and again this is not related to adrenaline release.[13]

It is likely that mild stress associated with experimental procedures also has effects on a variety of other physiological parameters and that these effects are not confined to sheep. Such information is sparse for the sheep but is also sparse for other laboratory animals. Such effects should be kept in mind when planning experiments. If the nature of the experiment requires a "normal" animal, time should be taken to accustom the animal to the laboratory and the laboratory routine. Sufficient time should elapse after any unpleasant procedures, such as insertion of a jugular cannula, for the effects to wear off before the experiment is performed. No strict rules can be laid down but a sheep should be considered accustomed to the laboratory conditions and should react minimally to procedures when it will take food from the hand. After moving to a new environment, about a week should elapse before experiments are started and any stressful procedures such as cannulation should be done on the day prior to the experiment.

4.4 Other Aspects of Behaviour

Although little sheep research has involved conditioned reflexes or operant conditioning, both of these experimental tools can be used. Sheep have good memories for routes through sheep yards and can learn a route in a relatively small number of trials.[21] An interesting conditioned reflex is that involving the oesophageal groove (p. 24) where merely showing a bottle to an adult sheep in which the groove-closing reflex has been maintained will cause swallowed fluids to pass directly into the abomasum.[35] Operant conditioning has been used to determine illumination preferences for sheep[5] and to monitor effects of chronic lead ingestion.[16]

Administration of sex steroids to ovariectomized ewes produces some interesting results. Single injections of either oestradiol-17β or testosterone induces oestrous behaviour[8] but chronic treatment with either of these hormones causes male-like behaviour.[44] If ewes are "androgenized" *in utero* by administration of testosterone to the ewe, ewe lambs after puberty will show permanent male sex behaviour.[8] These may even obviate the necessity for vasectomized rams.[28] Castrate male sheep (wethers) treated with either oestrogen or testosterone for a week also show typical male behaviour and can detect ewes in oestrus well. Such animals might have a use as "teasers"[15] (see p. 145).

References

1. Alexander, G. (1980). Husbandry practices in relation to maternal offspring behaviour, pp. 99–107. (In reference 51.)
2. Arnold, G. W. and Dudzinski, M. L. (1978). *Ethology of Free-ranging Domestic Animals*. Elsevier Scientific Publ. Co.; Amsterdam.
3. Arnold, G. W. and Maller, R. A. (1974). Some aspects of competition between sheep for supplementary food. *Anim. Prod.*, **19**: 309–319.
4. Arnold, G. W. and Pahl, P. J. (1974). Some aspects of social behaviour in domestic sheep. *Anim. Behav.*, **22**: 592–600.
5. Baldwin, B. A. and Start, I. B. (1978). Methods for the study of illumination preferences in sheep and calves. *J. Physiol.*, **284**: 13P–14P.
6. Bray, A. R. and Wodzicka-Tomaszewska, M. (1974). Perinatal behaviour and progesterone and corticosteroid levels in sheep. *Proc. Aust. Soc. anim. Prod.*, **10**: 318–321.
7. Caple, I. W. and Heath, T. J. (1978). Regulation of excretion copper in bile of sheep: Effect of anaesthesia and surgery. *Comp. Biochem. Physiol.*, **61A**: 503–507.
8. Clarke, I. J. and Scaramuzzi, R. J. (1978). Sexual behaviour and LH secretion in spayed androgenized ewes after a single injection of testosterone or oestradiol 17β. *J. Reprod. Fertil.*, **52**: 313–320. (See also Clarke, I. J. (1978/1979). *Anim. Reprod. Sci.*, **1**: 305–312.)
9. Done, J. R. (1975). *Ethological Aspects of Sheep in an Unnatural Environment, the Animal House*. Thesis for B.R.Sc(Hons), University of New England, Armidale, Australia.
10. Dove, H., Beilharz, R. G. and Black, J. L. (1974). Dominance patterns and positional behaviour of sheep in yards. *Anim. Prod.*, **19**: 157–168.
11. Dulphy, J. P., Remond, B. and Theriez, M. (1980). Ingestive behaviour and related activities in ruminants, pp. 103–122. (In reference 313 in Chapter 3.)
12. Espiner, E. A., Hart, D. S. and Beaven, D. W. (1972). Cortisol secretion

during acute stress and response to dexamethasone in sheep with adrenal transplants. *Endocrinology*, **90**: 1510–1514.

13. Falconer, I. R. and Jacks, F. (1975). Effect of adrenal hormones on thyroid secretion and thyroid hormones on adrenal secretion in the sheep. *J. Physiol.*, **250**: 261–273.
14. Fletcher, I. C. and Lindsay, D. R. (1968). Sensory involvement in the mating behaviour of domestic sheep. *Anim. Behav.*, **16**: 410–414.
15. Fulkerson, W. J., Adams, N. R. and Gherardi, P. B. (1981). Ability of castrate male sheep treated with oestrogen or testosterone to induce and detect oestrus in ewes. *Appl. anim. Ethology*, **7**: 57–66.
16. Gelder, G. A. van, Carson, T., Smith, R. M. and Buck, W. B. (1973). Behavioural toxicologic assessment of the neurologic effect of lead in sheep. *Clin. Toxicol.*, **6**: 405–418.
17. Gohary, G. S. and Bickhardt, K. (1979). Der Einfluss des Blutentnahmestresses auf Blutmesswerte des Schafes. *Deutsch. Tier. Wochen.*, **86**: 225–228.
18. Hulet, C. V., Alexander, G. and Hafez, E. S. E. (1975). The behaviour of sheep, pp. 246–294. In *The Behaviour of Domestic Animals*, 3rd edition (edited by Hafez, E. S. E.). Bailliere Tindall; London.
19. Holley, D. C. *et al.* (1975). (See reference 189 in Chapter 3.)
21. Hutson, G. D. (1980). The effect of previous experience on sheep movement through yards. *Appl. anim. Ethology.*, **6**: 233–240.
22. Key, C. and MacIver, R. M. (1980). The effects of maternal influences on sheep: Breed differences in grazing, resting and courtship behaviour. *Appl. anim. Ethology.*, **6**: 33–48.
23. Kilgour, R. and DeLangen, H. (1970). Stress in sheep resulting from management practices. *Proc. N.Z. Soc. anim. Prod.*, **30**: 65–76.
24. Knight, T. W. and Lynch, P. R. (1980). Source of ram pheromones that stimulate ovulation in the ewe. *Anim. Reprod. Sci.*, **3**: 133–136.
25. Lees, J. L. and Weatherhead, M. (1970). A note on mating preferences of Clun Forest ewes. *Anim. Prod.*, **12**: 173–175.
26. Lindsay, D. R. and Fletcher, I. C. (1972). Ram-seeking activity associated with oestrous behaviour in ewes. *Anim. Behaviour*, **20**: 452–456.
27. Lynch, J. J., Brown, G. D., May, P. F. and Donnelly, J. B. (1972). The effect of withholding drinking water on wool growth and lamb production of grazing Merino sheep in a temperate climate. *Aust. J. agric. Res.*, **23**: 659–668.
28. Marit, G. B., Scheffrahn, N. S., Troxel, T. R. and Kesler, D. J. (1979). Sex behavior and hormone responses in ewes administered testosterone propionate. *Theriogenology*, **12**: 375–381.
29. Mattner, P. E. and Braden, A. W. H. (1975). Studies of flock mating of sheep: 6. Influence of age, hormone treatment, shearing and diet on the libido of Merino rams. *Aust. J. exp. Agric. anim. Husb.*, **15**: 330–336.
30. Mattner, P. E., George, J. M. and Braden, A. W. H. (1976). Testosterone treatment of ram lambs: Effect on adult libido. *Theriogenology*, **6**: 613.
31. McNatty, K. P. and Thurley, D. C. (1973). (See reference 256 in Chapter 3.)

32. Morag, M. (1967). Influence of diet on the behaviour pattern of sheep. *Nature*, **213**: 110.
33. Morgan, P. D., Boundy, C. A. P., Arnold, G. W. and Lindsay, D. R. (1975). The roles played by the senses of the ewe in the location and recognition of lambs. *Appl. anim. Ethology*, **1**: 139–150.
34. Neathery, M. W. (1971). Acceptance of orphan lambs by tranquillized ewes (*Ovis aries*). *Anim. Behaviour*, **19**: 75–79.
35. Orskov, E. R. (1975). Physiological conditioning in ruminants. *World Animal Rev.*, **16**: 31–36.
36. Pearce, G. R. (1965). Rumination in sheep: The circadian pattern of rumination. *Aust. J. agric. Res.*, **16**: 635–648.
37. Pearson, R. A. and Mellor, D. J. (1976). Some behavioural and physiological changes in pregnant goats and sheep during adaption to laboratory conditions. *Res. vet. Sci.*, **20**: 215–217.
38. Poindron, P. and Carrick, M. J. (1976). Hearing recognition of the lamb by its mother. *Anim. Behaviour*, **24**: 600–602.
39. Rayssiguier, Y. (1977). Hypomagnesemia resulting from adrenaline infusion in ewes: Its relation to lipolysis. *Horm. Metab. Res.*, **9**: 309–314.
40. Reilly, P. E. B. and Chandrasena, L. G. (1979). Indices of stress in housed experimental sheep. *Res. vet. Sci.*, **27**: 250–252.
41. Ruckebusch, Y. (1972). The relevance of drowsiness in the circadian cycle of farm animals. *Anim. Behaviour*, **20**: 637–643.
42. Setchell, B. P. and Waites, G. M. H. (1962). Adrenaline release during insulin hypoglycaemia in the sheep. *J. Physiol.*, **164**: 200–209.
43. Shannon, A. D., Quin, J. W. and Jones, M. A. S. (1976). Response of the regional lymphatic system of the sheep to acute stress and adrenaline. *Quart. J. exp. Physiol.*, **61**: 169–184.
44. Signoret, J. P. (1980). Endocrine basis of reproductive behaviour in female domestic animals, pp. 1–9. (In reference 51.)
45. Smith, F. V., Van-Toller, C. and Boyes, T. (1966). The "critical period" in the attachment of lambs and ewes. *Anim. Behav.*, **14**: 120–125.
46. Trenkle, A. (1978). Relation of hormonal variations to nutritional studies and metabolism of ruminants. *J. Dairy Sci.*, **61**: 281–293.
47. Welch, R. A. S. and Kilgour, R. (1970). Mismothering among Romneys. *N.Z.J. Agric.*, **121**: 26–27.
48. Williams, H. Ll. (1978). Sheep, pp. 62–111. In *The Care and Management of Farm Animals* (edited by Scott, W. N. and Laing, J. A.). Bailliere Tindall; London.
49. Wilson, S., MacRae, J. C. and Buttery, P. J. (1979). The effects of implanting jugular catheters on plasma glucose concentrations in wethers fed two extremes of diet. *Res. vet. Sci.*, **26**: 256–258.
50. Winfield, C. G. and Makin, A. W. (1975). Prediction of the onset of parturition in sheep from observations of rectal temperature changes. *Livestock Prod. Sci.*, **2**: 393–399.
51. Wodzicka-Tomaszewska, M., Edey, T. N. and Lynch, J. J (1980), editors. *Behaviour in Relation to Reproduction, Management and Welfare of Farm Animals.* Reviews in Rural Science, No. 4, published by The

University of New England Publishing Unit, Armidale, Australia.
52. Wodzicka-Tomaszewska, M., Kilgour, R. and Ryan, M. (1981). Libido in the larger farm animals: A review. *Appl. anim. Ethology*, **7**: 203–238.
53. Zenchak, J. J., Anderson, G. C. and Schein, M. W. (1981). Sexual partner preference of adult rams (*Ovis aries*) as affected by social experiences during rearing. *Appl. anim. Ethology*, **7**: 157–167.

Chapter 5

Breeds and Supply of Sheep

5.1 Breeds

The number of breeds of sheep in the different sheep raising countries of the world is multitudinous. In Great Britain, the British Sheep Association recognizes 52 common breeds and six rare breeds.[14] Many of the breeds are similar and sometimes the same breed is given different names in different countries. The origins of many of the sheep breeds and their uses in different countries are given by Ryder and Stephenson.[28] In many countries in which sheep research is done, British breeds are used. Readers who are unfamiliar with the agricultural uses of sheep may therefore find the following greatly simplified classification useful.

Breeds of sheep have been developed with a greater or lesser emphasis on wool production depending on their use as meat animals and in a few countries for milk production. In general, the hill breeds (Table 37) are principally wool growing sheep which are used for mutton production and for breeding fat lamb mothers. In contrast, the short wool breeds are good meat producing animals and are used as sires of fat lambs but their wool is coarser and less valuable. Long-wool breeds produce considerable wool and are good meat producers.

In Great Britain and many other countries, the most usual usage of sheep is for the poorest land which tends to be high hill country and moors to be used for ewes and lambs of the hardy hill breeds. Most of the male lamb progeny are castrated and after weaning are sold for eventual fattening on better farming land as mutton animals. These hill breeds which comprise nearly 45% of British sheep tend to have

Table 37 A classification of some sheep breeds and their wool characteristics.

Type of shcep	Typical breeds	Wool characteritics "Cut" (kg/year)	Length (cm)	Count
Blackfaced, horned hill breeds	Scottish Blackface	2·0	23–35	28–40
Whitefaced, horned hill breeds	Cheviot, Welsh Mountain	2·1	10	50–56
Whitefaced short wools	Dorset Horn	3·0	8·6	50–56
Blackfaced short wools	Southdown	1·8	3·8–7·6	55–60
Demilustre medium long wools	Border Leicester	3·6	20–25	46–48
Lustre medium long wools	Leicester, Lincoln	5–6·4	23–35	36–46
Fine wools	Merino	6·4		60–90

(Adapted from [28].)

only one lamb per ewe per year. Surplus hill breed ewes are sold to "upland" farms where they are mated with long wool rams to produce lambs of which the males are castrated and fattened to produce a better quality mutton animal and the ewes are sold as fat lamb mothers. Both the meat quality and the fertility of these fat lamb mothers are higher than the hill breeds. The ewe produced by crossing the Scottish Blackface with the Border Leicester is called the Greyface while that produced by crossing the Welsh Mountain with the Border Leicester is called the Welsh Halfbred and by crossing the Cheviot with the Border Leicester the Scottish Halfbred.

These crossbred ewes are usually mated with a ram of one of the Down breeds (in some regions, the Dorset Horn or Polled Dorset breeds are used instead) and all progeny are fattened and sold within 6 months (ideally 3 to 4 months) as fat lambs. Some breeds do not fit into the above scheme, one being the Clun Forest which is found near the Welsh border and is used in several research institutes. The Cheviot is another. Both breeds are sometimes classed as "self-contained" or semi-hill breeds and both are often used directly as fat lamb mothers.

In some countries such as Australia, hill breed sheep are rare, but Merinos are common in the poorer country and are used to produce fat lamb mothers by mating to long wool rams. However, as wool production from Merinos is more important than fat lamb production, most Merino ewes are mated with Merino rams.

For wool production, the most important features are the "cut" (lb or kg of wool produced per year) and the fineness. Count (Bradford count) is a measure of fineness and theoretically is the number of hanks of wool each 560 yards long which can be spun from 1 lb of clean scoured wool. Typical values for count are given in Table 37. Count is being replaced as a measure by mean fibre diameter. Coarse wools (32 to 44 count) are used for carpets, medium wools (44 to 58 count) for knitting wools and fine wools (60 or greater count) for fine woven fabrics. The finest wools are produced by the Merino which tends to be grazed over vast areas in arid regions of countries such as Australia and South Africa. Length of wool is less important than fineness although for a given fineness, longer fibres are preferred. Crimp, the number of waves in the wool per inch, is approximately correlated with fineness as coarser wools may have as few as three crimps while Merino wools have up to 30.

The Merino is the most numerous breed of sheep in the world. It originated in Spain from where in the seventeenth century some animals were taken to France, Britain and Germany. Two strains persisted; the Rambouillet in France and the Saxony strain in Germany. These strains had an important role in improving strains descended from the original Merino in countries such as the United States, Australia and South Africa and in the United States, one strain of Merino is now known as the Rambouillet. Other Merino strains are known by various names such as the Arles in France.

In Australia, strong genetic selection applied to the Merino nearly tripled the average production of wool per sheep over a century. The Australian Merino has actually evolved into three strains. That with the coarsest wool (count 60–64) is large framed (ewes typically 45 kg) and is predominant in the drier parts of South Australia while the finest woolled strain (66–90 count) is of small body weight (ewes typically 30–35 kg) and is found in the cooler regions of the east coast and Tasmania. The third strain, sometimes known as the Peppin strain after one of the most important studs in its formation is the most common and predominates in the wheat growing areas.

Some breeds have been developed from the Merino in an attempt to improve its poor meat qualities and yet retain much of its excellent wool quality. The Corriedale was developed in New Zealand by crossing Merinos with the Lincoln. Corriedales have been exported to several countries including the United States. Similarly in the United States, the Columbia was developed by crossing Merinos with the Lincoln. Also in the United States, the Targhee was developed by backcrossing the Columbia to the Merino and thus is analogous to the Polworth which was developed in Australia by crossing the Corriedale with the Merino.

The influence of British breeds throughout the world has been considerable as Britain started farming pure "improved" breeds of sheep at the time that her later colonies were settled and many were exported to these colonies. They were also exported to many other countries for improving native breeds of sheep. In developing countries therefore, many native breeds have British breeds in their ancestry. The movement has not been all in one direction as is illustrated by the Polled Dorset breed which in Britain is rapidly replacing the Dorset Horn breed. The Polled Dorset was developed in Australia by selective breeding in order to eliminate the very large and weighty horns of the Dorset Horn.

The Finnish Landrace is a breed which has been widely used in several countries for cross breeding as it has good meat qualities and is also highly prolific. In this context, a strain of fine woolled Merino in Australia, the Booroola, deserves mention as it also has a high proportion of multiple births (up to five lambs). The physiological mechanisms causing multiple births in these two types of sheep have not been fully elucidated.

One rare breed also deserves mention as it has been used many times for research in Britain. It is the Soay which is a nearly pure descendant of sheep introduced to Britain by the Romans and which is still found on some of the Outer Hebridean islands in a feral state. It is very small (about 20 kg) and so corresponds to the miniature pig which is preferred to normal pigs in some research institutes. The wool of the Soay is brown and each year it moults and so does not have to be shorn.

Some typical values for body weight and lambing percentage are given in Table 38.

Table 38 Typical weights, fleece weights and lambing percentages of ewes of different breeds.

Breed	Live weight of mature ewe kg	Lambing percentage normal range	Fleece weight kg
Cheviot	59–73	141–158	1·8–2·3
Clun Forest	56	150–164	2·5
Corriedale	34–39	110–114	3·5–4·0
Dorset Horn	73	127–158	2·7
Finnish Landrace	50	200–400	1·0–3·0
Lincoln	91	129–157	6·4
Merino	45	103–161	6·7
Welsh Mountain	30	95	1·0–1·1

(From [30].)

5.2 Supply

Some research institutes select particular types of sheep for experiments. Reasons for the selection are:

(1) **Size.** For some experiments such as those involving implantation of artificial devices, large sheep are needed. The Suffolk may be selected as its body weight can be greater than that of a man. For most other experiments, smaller sheep are preferable, being easier to lift, requiring smaller pens and less food and usually being cheaper. The Soay in particular and the Merino and Welsh Mountain being smaller breeds are better for X-radiography.

(2) **Number of Lambs.** For foetal research, some experiments require single pregnancies while others require twins. Requirements may also vary for reproduction experiments.

(3) **Age.** It is a common opinion that young sheep are more resistant to effects of major surgery than are older sheep.

(4) **Behaviour.** Some researchers avoid the hill breed sheep as they believe that sheep of these breeds are more independent in their behaviour and hence do not settle down as well in the labora-

tory. However it is the experience of others that sheep of the hill breeds are very suitable as laboratory animals and that they do not have behavioural problems in the laboratory.

In the absence of preference factors such as the above, breeds tend to be selected either on the basis of cost or because previous experiments in a laboratory have been done with a particular breed.

There is little in the literature on how individual sheep differ physiologically in other than traits with obvious genetic association. Some differences which have been described are:

Excretion of ammonia in urine;[30]
Mitogen-induced transformation of lymphocytes;[10]
Rumen volume—probably associated with rate of eating, dry matter digestibility and ammonia utilization.[19]

Sources of supply of sheep for experimentation are three—a research institute farm, a cooperative farmer or a livestock market. Before discussing these sources, it is worthwhile considering which types of sheep are likely to be cheapest as cost is a factor in many experiments. In general, it will be cheaper to buy sheep "off shears" (i.e. recently shorn). Even if "off shears" sheep are not more economical, it may still be advisable to buy them as control of external parasites is only feasible in recently shorn sheep (p. 202). Store sheep (sheep in lean rather than in fat condition) are always cheaper because fat sheep are destined for slaughter whereas store sheep require prior fattening. Hill breeds and the Merino are usually cheaper than other types of sheep as their meat quality is poorer when fattened than that of the usual meat breeds. Rams are available only in small numbers and castrated males (wethers) are normally sold in fat condition. Finally, older ewes will be cheaper than young maiden ewes due to a reduced potential for bearing lambs. From these considerations, the cheapest type of sheep is likely to be a ewe of a hill breed or Merino type which has been sold after rearing several lambs. This generalization may be influenced by local considerations such as local scarcity of sheep of particular types and the type of experiment.

Sheep have a potential lifespan of 15 to 20 years in the laboratory. On farms, they are kept for a much shorter period and are usually sold at between 5 and 8 years. Such sheep are usually called "culls" and hence a sheep culled from a farm will still have a potential lifespan of several years when kept under good conditions such as are likely to exist in a laboratory animal house.

The reasons why sheep are culled from farms are numerous. Sheep at pasture do not fare well when they lose one or more of the incisor teeth (become "broken mouthed") and a broken-mouthed sheep will be culled by a farmer as soon as is practical although one study in England showed that the body condition of culled ewes was not adversely affected by dental disease.[25] Little study has been done on the causes of loss of teeth in sheep but the indications are that there are close parallels with human periodontal disease. In many sheep there is a chronic gingivitis associated with accumulation of subgingival plaque. Before teeth are lost, the gingivitis and plaque accumulation become severe and lead to loss of connective tissue and disorganization of fibre groups of the periodontal ligament.[31] The only palliative is the fitting of dentures[15] but this is not necessary for housed sheep and is not practical on farms. Likewise a ewe which has lost one side of the udder from mastitis or due to a teat being cut off by a careless shearer is liable to be culled, especially if she is of a breed that has multiple births. This defect may not affect her use in the laboratory. Apart from these specific reasons for culling, wool growth and fertility decrease slightly after about 5 years and mortality in a flock from natural causes will increase after this time. A survey of cull sheep bought at a market in England indicated that most had a combination of foot troubles, affected udders and loss of one or more teeth. A few were also culled for infertility.[8] However infertile sheep are likely to become fat and hence will probably be sold for slaughter.

Thus the cheapest sheep for research purposes are likely to be ewes of the Merino or hill breeds which have been culled. It must be stressed that although they will probably have been culled for age-related reasons, they are not old in a biological sense. Provided that they are not excessively thin (implying some disease), they should be suitable for many areas of research and are likely to be much healthier than mongrel dogs which are frequently used as research animals. The loss of incisor teeth or of half of an udder is not likely to affect their viability in the laboratory.

The most convenient source of supply of sheep for research is an institute farm. Most of the Agricultural Research Institutes and a number of the Universities in Britain, United States, Australia and New Zealand have associated farms which supply many of the sheep used for agricultural-type research. However, few sheep for medical-type research come from such farms. Ideally a research institute farm should be sufficiently large that it can be run as a cost-recovery enterprise which produces an adequate number of surplus animals for research. Scientists often have difficulty in predicting several months in advance what their animal requirements are likely to be but those not needed can be sold if the farm is geared to producing a surplus to anticipated requirements. Although it is usually necessary to recover most or all of the costs, such research farm enterprises must be run with the interests of research kept foremost and profit must not be a prime motive.

Sheep raising tends not to be a labour intensive process when compared with raising pigs, dogs or small laboratory animals for research. There are several books written on commercial sheep raising and these should be consulted for details.

Some research institutes arrange with a cooperative nearby farmer to supply a set number of surplus sheep each year at a contract price. Although it is not often done, better cooperation will be obtained if the farmer is invited to the laboratory to see what is being done and for the importance of the research projects to be explained. A cooperative farmer may be able to supply a set number of ewes each year with known gestation dates (see below) or to keep a number of sheep on agistment.

Many sheep for research are purchased on the open market. A livestock agent should be selected and made aware of the type of sheep that will be required; their age, sex, condition, breed etc. He will know what the prices and likely market trends are and at what time of the year they will be cheapest or most plentiful. If needs for several months can be predicted, it may be worthwhile buying animals in reasonable numbers infrequently rather than in small numbers every week or two; the excess to immediate needs being kept on agistment. An agent will also be able to sell animals after the experiment, provided that the nature of the experiment and local animal experimentation regulations permit such sale.

5.3 Introduction to the Animal House

Unless sheep are known to come from a lice or a ked free farm, and have not become contaminated during transport, they should be sprayed with a suitable insecticide to kill lice or ked before introduction to the animal house. A high pressure spray is needed to penetrate the wool unless sheep have been shorn recently. In the field, the insecticidal effects of such spraying decreases with time due to leaching by rain but in the laboratory spraying should have a longer term effect as sheep are rarely wet. Quarantine for 24 to 48 h may be needed to allow the insecticide to act so as to avoid infesting sheep already in the animal house. In practice, sheep introduced to a laboratory animal house are unlikely to be incubating contagious diseases which will become obvious during a normal quarantine period of 2 to 3 weeks. Therefore there are no strong reasons for quarantine of newly introduced sheep for longer than a day or two.

On introduction, the following features should be inspected:

(1) **Feet.** These may need cleaning or trimming (p. 161).
(2) **Teeth.** Looseness or loss of incisor teeth should be noted. The molar and premolar teeth are difficult to inspect and the state of the incisor provides a poor indication of the condition of the cheek teeth.[25]
(3) **Udder.** An undeveloped udder in a ewe is a sign that she has never had a lamb while a pronounced asymmetry or a lump indicates previous mastitis.
(4) **Testes.** Asymmetry is suggestive of epididymitis (p. 199) but can be due to other conditions such as a varicocoele.[18,38]
(5) **Infected lymph nodes** (p. 197).
(6) **Body condition.** This can be deceptive in a sheep with long wool. It is best assessed by palpating the relative prominence of the transverse and dorsal processes of the lumbar vertebrae.[40]
(7) **Chronic cough.** Sheep should be allowed to settle down before listening for a cough. If several sheep are introduced at one time, any that have an unusually rapid respiration rate (provided that the wool covering is comparable between sheep) may be suspected of having a lung lesion (p. 197).

A blood examination may be performed for anaemia but this is only warranted for an animal in poor condition. Paleness of mucous membranes (e.g. under the lower eye lid) gives a good indication of anaemia. Such animals should be inspected carefully to determine if there is an obvious cause for the poor condition. A faecal worm egg count may also be done but it is better to dose routinely every sheep that is introduced to the animal house with suitable anthelmintics. A dose of one of the newer benzimidazole drugs (e.g. Fenbendazole, Albendazole, Oxfendazole) followed 2 or 3 days later with Levamisole, each at twice the recommended dose rate, should kill most intestinal worms including tape worms, liver fluke and lung worms. These drugs have wide margins of safety and the recommended dose rates are based on economical control of parasites on farms. In the laboratory, cost of anthelmintics will be negligible and hence the dose rate can be increased. The combination is recommended as some parasites may have resistance to a particular anthelmintic. Several larval parasites are more resistant and the dosing could be repeated after a month to kill any larval stages that have matured.

After introduction to the animal house, new sheep should be watched for a few days to ensure that they become accustomed to a change of diet.

5.4 A Supply of Ewes with Known Gestation Dates

Many research projects involve the use of pregnant ewes with known gestation dates. Their supply can present a problem as such ewes cannot be purchased on the open market but must instead be "hand mated". This may be done by a cooperative farmer or on a research institute farm. The discussion presented below is based on a laboratory requiring a small number (say 2 to 4/week) of pregnant ewes with known gestational ages. It may be of help to read the section on reproductive physiology (p. 61) first to assist in understanding the rationale behind some of the suggestions.

Sheep are seasonal breeders and hand mating in the normal breeding season will be considered first. Without a ram being present, it is difficult to detect the occurrence of oestrus in ewes by their behaviour (p. 121). Changes in vaginal histology at different stages of the cycle are not as distinct as in animals such as the rat, especially in the posterior part of the vagina.[7] However there is a fairly uniform sequence of changes in the vaginal mucous at about the time of oestrus,[20] although this has rarely been used as a guide to occurrence of oestrus.

For detection of oestrus, usually a vasectomized ram is fitted with a harness holding a coloured crayon on the brisket[21] and is run with a flock of ewes. When mating is attempted, the crayon leaves marks on the wool of the rump of the ewe. Available colours of crayons are red, blue, green and yellow and they are made in hard, medium and soft consistencies for hot, neutral and cold conditions respectively. Colours should be changed every 14 days which is sufficiently short to avoid a ewe being marked twice with one colour in two successive oestrous periods. It is necessary for the flock to be inspected once or twice each day either by yarding or else walking quietly through the flock and noting ewes marked since the previous inspection. Numbers or letters sprayed on the flank provide easy identification of individual ewes. When the oestrous pattern of the flock has been ascertained, ewes can be removed from the flock as necessary for mating with a fertile ram.

If there is only a small number of ewes in an experiment, they may be run in a large pen or a small paddock. The services of a vasectomized ram are not required either if the fertile ram is put in the pen for about 5 min twice per day or the ewes are herded into a yard and the ram introduced. Once a ewe is identified by behaviour as being on oestrus, she can be removed from the group before she is served so that the ram can search out any other ewe on oestrus. If mating is required, she can then be put in the ram's pen for a few hours to be served several times. A group of sheep will rapidly become accustomed to the routine of yarding, introducing the ram and then removing the ram and after the first week the routine can be done in less than 10 min. It is not necessary to cover the rams abdomen to prevent mating as ewes on oestrus can quickly be separated from the ram once they have shown behavioural signs.

Ewes on oestrus often seek out the ram and hence, if a ram is kept in a pen or small field next to that of the ewes, a ewe on oestrus can frequently be identified by her behaviour before the ram is introduced.

Although one ejaculation from a ram is theoretically sufficient to result in conception, the pregnancy rate is higher if a ram mates with a ewe several times. Hence the ewe should be left with the ram for a few hours. Losses post-conception have been estimated to be 20–30% in the first 30 days[6] and losses after this time are small unless there is venereal disease in the flock.

A general review of methods for pregnancy diagnosis in the ewe is given by Richardson.[22] A pregnancy test (rosette inhibition test) has recently been developed which with an occasional false negative can detect pregnancy as early as 1 to 3 days of mating[17] but it does not work after 4 weeks of pregnancy. As HCG-like hormones are not produced in pregnant sheep, there is no test similar to the currently used human pregnancy tests. Apart from the rosette inhibition test, the simplest methods for diagnosing pregnancy are either to look for oestrous behaviour 16 to 19 days after mating or to take samples of blood on days 10 and 17 after mating. Progesterone should be present at greater than 1 ng/ml in both samples if a ewe is pregnant but should be negligible in the second sample if she is not. If the mating dates of ewes are not known, then three samples should be taken, on days 1, 6 and 12.[39] A low level (<1 ng/ml) in one is suggestive that a ewe is not pregnant.

A very occasional ewe will show oestrous behaviour when pregnant. One report gave the incidence of "false" oestrus at 22% with a mean interval of 21·5 days after conception for occurrence[41] but it is possible that such ewes were initially pregnant but later lost the embryo (see above). In the author's experience, false oestrus is uncommon but may occasionally be seen up to 70 days after mating.

Pregnancy diagnosis may be required after implantation has occurred. Assay of ovine chorionic somatomammotrophin after day 55 is 99% accurate.[26] Radiographic examination is accurate for diagnosis of pregnancy and for foetal numbers with a detection level of over 90% after 70 days and over 99% after 81 days.[39] For radiography, a strong compression band used on ewes fasted for 24 h gives the best results.[39] A small piece of cotton wool dampened with ether and placed over the ewe's nostrils will cause her momentarily to stop breathing, allowing the X-ray exposure to be made without blurring from respiratory movements.[2]

Foetal heart beats can be detected by Doppler probes or microphones from about 40 days after mating but are easier to detect from older foetuses.[5,12] The probe may be inserted into the rectum or may be of the pencil type placed on the midline of the lower abdomen. The ewe is best placed on her rump for the latter procedure.

Hulet[9] suggested a rectal probe method for pregnancy diagnosis. The sheep was sat on its rump and the foetus was "felt" with the probe. This method is not recommended as, although it is accurate, a proportion of sheep may abort or die.[35]

It may be necesary to confirm foetal ages from crown-rump

lengths. Several formulae have been determined for sheep:

Y (days) $= (X + 164)/4{\cdot}87$ cm, where X is the crown-rump length.[23]

Y (cm) $= (-0{\cdot}28 - 4{\cdot}56\,X + 1{\cdot}45\,X^2)$ where $X =$ days/100.[33]

Also tibial and radial lengths may be used[24] as can the appearance of ossification centres.[11,24]

In spite of the relatively constant gestation length of ewes, it is difficult to predict just when a ewe is likely to give birth until parturition has actually started. The udder increases in size at a variable time before birth and relaxation of the vulva is gradual. If it is necessary for a person to be present at birth and the mating date is known, birth my be induced with corticosteroids just before the anticipated birth date (p. 66).

It may be that many ewes in oestrus are required on a particular day.[13] In a large flock, 5 or 6 out of 100 ewes on average can be expected to be in oestrus each day. Two methods are available for synchronizing oestrus in order to increase the number on a particular day. A progestagen impregnated sponge, pessary or tampon can be inserted into the vagina of each ewe.[42] The usual recommendation is that it be left in for 14 days[27] but experiments in the author's laboratory indicate that a period of 10 days gives the same degree of synchronization and higher fertility.[4] Subcutaneous implants in the ear can also be used.[32] Most ewes will show oestrous behaviour about 40 h after removal of the tampon (range 24 to 120 h) but a proportion will not become pregnant. If ewes are not mated at the oestrus following removal, the subsequent oestrus will have higher fertility but less synchronization. The second method involves an intramuscular injection of an analogue of prostaglandin $F_{2\alpha}$ (Cloprostenol, 100 ug) to destroy the corpus luteum.[34] Eighty-four per cent of ewes will be on oestrus after 28 to 48 h. As a proportion of the injected ewes will not have a functional corpus luteum (i.e. are approaching oestrus or have recently been on oestrus and so will have an immature corpus luteum which is refractory to the drug), it is better to give a second injection 10 days later when all corpora lutea will be susceptible and so all ewes should show oestrous behaviour.

While 70% or more of non-synchronized ewes should lamb after a first service with a ram, the percentage will be lower for ewes synchronized with progestagen sponges and lower still when prostaglandins are used.[1] Allowance should be made for this lower fertility when determining the number of ewes to mate. If a definite number of ewes is required on a particular day, it could be worthwhile mating a larger number and then aborting the surplus with Cloprostenol after about 40 days (i.e. after the elapse of two oestrous cycles).

It can be difficult to get ewes pregnant when it is out of the normal breeding season (autumn and winter). Two approaches have been tried. Hormone treatment has been given to stimulate the ovary and induce oestrus. Alternately, the daylight pattern has been altered by housing the ewes indoors for some 4 to 8 weeks and reducing the period of illumination by about half an hour each week. For each of these approaches, the results have been rather unsatisfactory as half or fewer ewes become pregnant with the remainder relapsing into anoestrus. To obtain lambs throughout the year for experiments, Mears *et al.*[16] combined these two approaches and obtained a conception rate of 80%. In their protocol, they housed the ewes indoors and gave them a shorter period of illumination than they had been receiving in the field (10 h light per 24 h) and kept the temperature constant (18–20°C). After introduction to the animal house, each ewe received 14 daily injections of progesterone (7·5 or 10 mg/day). Then on day 15, 500 i.u. of PMSG was given and on day 16, 30 ug of oestradiol-17β. At the oestrus that followed the oestradiol injection, ewes were mated and returned to the paddock.

For sheep bred out of season, progesterone levels in blood should be taken for early diagnosis of pregnancy (p. 146). Batches of PMSG vary in potency and the dose may need to be altered with each new batch that is used.

The suggestions given above for obtaining pregnant ewes are mostly well-tried and should be effective. However, as with everything, there can be hidden traps. For instance, an occasional vaginal tampon may be lost. Also a ewe will very occasionally be marked by a ram with a harness and crayon when not in oestrus although this usually occurs on the first day or two after a fresh ram is introduced to a flock of ewes. A more serious problem would be infertility of the ram. On a farm, a flock of ewes will usually be sufficiently large that several rams are needed and if one happens to be infertile or to lack libido the other rams will be able to cover most ewes. However if one ram purchased for a small group of ewes is inadequate, then the consequences are serious. The first indication would be that most or all of a group of ewes mated with the ram would return to oestrus 16 to 18 days later. As a precaution, the fertility of such a single ram could be checked when he is first purchased, either by veterinarian or by test mating with a small number of ewes which could then be aborted with Cloprostenal once the ram's fertility has been proven. This treatment has little effect of subsequent fertility.[37]

To check out a ram, the testes should be palpated for bilateral symmetry and for unusual lumps which might be a sign of epididymitis.[18] Also a sample of semen should be obtained by electrical stimulation of a bipolar electrode inserted into the rectum[3] but if such a stimulator is not available then a sample of washings of the vagina of a ewe taken immediately after mating will give an indication of semen quality.

References

1. Allison, A. J. and Kelly, R. W. (1978). Synchronization of oestrus and fertility in sheep treated with progestagen-impregnated implants, and prostaglandins with or without intravaginal sponges and subcutaneous pregnant mare's serum. *N.Z. J. agric. Res.*, **21**: 389–393.
2. Ardran, G. M. and Brown, T. H. (1964). X-ray diagnosis of pregnancy in sheep with special reference to the determination of the number of foetuses. *J. agric. Sci.*, **63**: 205–207.
3. Blackshaw, A. W. (1954). A bipolar rectal electrode for the electrical production of ejaculation in sheep. *Aust. vet. J.*, **30**: 249–250.
4. Bray, A. (1978). (See reference 53 in Chapter 3.)
5. Deas, D. W. (1977). Pregnancy diagnosis in the ewe by an ultrasonic rectal probe. *Vet. Rec.*, **101**: 113–115.
6. Edey, T. N. (1969). Prenatal mortality in sheep: A review. *Anim. Breeding Abstr.*, **37**: 173–190.
7. Ghannam, S. A. M., Bosc, M. J. and du Mesnil du Buisson, F. (1972). Examination of vaginal epithelium of the sheep and its use in pregnancy diagnosis. *Am. J. vet. Res.*, **33**: 1175–1185.
8. Herrtage, M. E., Saunders, R. W. and Terlecki, S. (1974). Physical examination of cull ewes at point of slaughter. *Vet. Rec.*, **105**: 257–260.
9. Hulet, C. V. (1975). Determining fetal numbers in pregnant ewes. *J. anim. Sci.*, **36**: 325–330.
10. Larsen, H. J. (1979). A whole blood method for measuring mitogen-induced transformation of sheep lymphocytes. *Res. vet. Sci.*, **27**: 334–338.
11. Lascelles, A. K. (1959). The time of appearance of ossification centres in the peppin-type Merino. *Aust. J. Zool.*, **7**: 79–86.

12. Lindahl, I. L. (1971). Pregnancy diagnosis in the ewe by intrarectal doppler. *J. anim. Sci.*, **32**: 922–925.
13. Love, J. A. (1978). Controlled breeding of sheep for studies in fetal physiology. *Lab. anim. Sci.*, **28**: 611–614.
14. McDougall, D. S. A. (1976), editor. *British Sheep*, 4th edition. The National Sheep Association.
15. Markham, J. H. A. and Lyle Stewart, W. (1962). Prosthetics in sheep. *Br. dental J.*, **112**: 327–329.
16. Mears, G. J., Van Petten, G. R., Harris, W. H., Bell, J. U. and Lorscheider, F. L. (1979). Induction of oestrus and fertility in the anoestrous ewe with hormones and controlled lighting and temperature. *J. Reprod. Fertil.*, **57**: 461–467.
17. Morton, H., Nancarrow, C. D., Scaramuzzi, R. J., Evison, B. M. and Clunie, G. J. A. (1979). Detection of early pregnancy in sheep by the rosette inhibition test. *J. Reprod. Fertil.*, **56**: 75–80.
18. Ott, R. S. and Memon, M. A. (1980). Breeding soundness examinations of rams and bucks, a review. *Theriogenology*, **13**: 155–164.
19. Purser, D. B. and Moir, R. J. (1966). Variations in rumen volume and associated factors influencing metabolism and protozoa concentrations in the rumen of sheep. *J. anim. Sci.*, **25**: 516–520.
20. Radford, H. M. and Watson, R. H. (1955). Changes in the vaginal contents of the Merino ewe throughout the year. *Aust. J. agric. Res.*, **6**: 431–445.
21. Radford, H. M., Watson, R. H. and Wood, G. F. (1960). A crayon and associated harness for the detection of mating under field conditions. *Aust. vet. J.*, **36**: 57–66.
22. Richardson, C. (1972). Pregnancy diagnosis in the ewe: A review. *Vet. Rec.*, **90**: 264–275.
23. Richardson, C. and Hebert, C. N. (1978). Growth rates and patterns of organs and tissues in the ovine foetus. *Br. vet. J.*, **134**: 181–189.
24. Richardson, C., Hebert, C. N. and Terlecki, S. (1976). Estimation of the developmental age of the ovine fetus and lamb. *Vet. Rec.*, **99**: 22–26.
25. Richardson, C., Richards, M., Terlecki, S. and Miller, W. M. (1979). Jaws of adult culled ewes. *J. agric. Sci.*, **93**: 521–529.
26. Robertson, H. A., Chan, J. S. D., Hackett, A. J., Marcus, G. J. and Friesen, H. G. (1980). Diagnosis of pregnancy in the ewe at mid-gestation. *Anim. Reprod. Sci.*, **3**: 69–71.
27. Robinson, T. J. (1967), editor. *The Control of the Ovarian Cycle in the Sheep*. Sydney University Press; Sydney, Australia.
28. Ryder, M. L. and Stephenson, S. K. (1968). (See reference 35 in Chapter 1.)
29. Scott, D. (1969). Renal excretion of potassium and acid by sheep. *Quart. J. exp. Physiol.*, **54**: 412–422.
30. Spedding, C. R. W. (1970), 2nd edition. *Sheep Production and Grazing Management*. Bailliere Tindall and Cassell: London.
31. Spence, J. A., Aitchison, G. U. and Sykes, A. R. (198). Broken mouth

(premature incisor loss) in sheep: The pathogenesis of periodontal diseases. *J. comp. Path.*, **90**: 275–292.

32. Spitzer, J. C. and Carpenter, R. H. (1979). Synchronized breeding of cycling ewes to produce fetuses of known gestational age. *Lab. anim. Sci.*, **29**: 755–758.
33. Thurley, D. C., Revfeim, K. J. A. and Wilson, D. A. (1973). Growth of the Romney sheep foetus. *N.Z. J. agric. Res.*, **16**: 111–114.
34. Trounson, A. O., Willadsen, S. M. and Moor, R. M. (1976). Effect of prostaglandin analogue Cloprostenol on oestrus, ovulation and embryonic viability in sheep. *J. agric. Sci.*, **86**: 609–611.
35. Turner, C. B. and Hindson, J. C. (1975) An assessment of a method of manual pregnancy diagnosis in the ewe. *Vet. Rec.*, **96**: 56–58.
36. Tyrrell, R. N., Gleeson, A. R., Peter, D. A. and Connell, P. J. (1980). Early identification of non-pregnant and pregnant ewes in the field using circulating progesterone concentrations. *Anim. Reprod. Sci.*, **3**: 149–153.
36. Tyrrell, R. N., Lane, J. G., Nancarrow, C. D. and Connell, P. J. (1981). Termination of early pregnancy in ewes by use of a prostaglandin analogue and subsequent fertility. *Aust. vet. J.*, **57**: 76–79.
38. Watt, D. A. (1978). Testicular pathology of Merino rams. *Aust. vet. J.*, **54**: 473–478.
39. Wenham, G. and Robinson, J. J. (1972). Radiographic pregnancy diagnosis in sheep. *J. agric. Sci.*, **78**: 233–238.
40. Williams, H. Ll. (1978). (See reference 48 in Chapter 4.)
41. Williams, S. M., Garrigus, U. S., Norton, H. W. and Nalbandov, A. V. (1956). The occurrence of estrus in pregnant ewes. *J. anim. Sci.*, **15**: 978–983.
42. Wishart, D. F. (1967). Synchronization of oestrus in sheep: The use of pessaries. *Vet. Rec.*, **81**: 276–287.

Chapter 6

Management of Experimental Sheep

6.1 Pens

A. *Sizes*

Sizes of pens used for experimental sheep vary considerably. Ideally, each animal should have ample room in which to move about in its pen. However, sheep are comparatively large experimental animals and space in most laboratory animal facilities is usually at a premium. Hence the desire is for pens for sheep to be as small as is compatible with reasonable comfort. The difficulty comes in defining what is reasonable comfort and thus how large pens should be. Recommendations for farm sheep housed indoors in group pens in winter provide a guide. These vary depending on the size of sheep and whether the flooring is slats or straw (Table 39) and seem to be reasonable for sheep in group pens in the laboratory. An increase of 50% in size for single pens also seems reasonable. Such sizes should allow sheep to turn around and to lie down with ease. In many laboratories, pens for sheep are smaller.

Table 39 Recommended sizes of group pens for sheep (from [34]).

Flooring	Slats	Straw
Large sheep (68–90 kg)	0·95–1·1 m^2	1·2–1·4 m^2
Small sheep (45–68 kg)	0·75–0·95	1·0–1·3
Hogget (32 kg)	0·55–0·75	0·75–0·95

B. *Floors*

Regulations for experimental animals usually require that floors of animal holding areas should be constructed with water-impervious materials. Such floors are often made from concrete coated with a sealant such as an epoxy resin. However, they are unsuitable for sheep, being smooth and not providing a grip for the feet so that animals have a tendency to slip. In most animal houses, sheep are either kept on an impervious floor which is strewn with straw or sawdust or are raised above it on a false floor of wooden slats or metal mesh. Straw is common in agricultural research institutes where it is usually readily available in the form of wheat stubble which has been baled. It acts in the pen to soak up urine and is replaced about once each week. It has several disadvantages—2 or more hundred kg may be needed for each sheep per year, considerable labour is needed for cleaning pens, straw will readily block drains and disposal of the mixture of straw and excreta in cities can be a problem.

With present high labour costs, slatted floors are more appealing. Sheep excreta are less objectionable than excreta from most other laboratory animals and several animal houses have raised sheep pens with spaces of about a metre beneath the pen floor. Provided that ventilation is good, the excreta are allowed to accumulate beneath the pen floor for several months before being removed. Such material is excellent for gardens and keen gardeners will readily take over the responsibility for its removal. There is comparatively little smell if ventilation is adequate and if there is not excessive spillage of water. An alternative system is to have the pen floor at least 70 cm above the floor of the room so that a high pressure hose can be used to wash accumulated excreta down a drain each day or two. The local authority should be consulted before this system is introduced. Also drains should be sufficiently large that they do not block with excreta and plenty of water should be used.

The design of the sheep housing area in the author's laboratory is such that the sheep pens are at floor level but under these pens there is a corrugated concrete floor sloping to a drain at an angle of 13°. At the top of the sloping floor is a perforated pipe running lengthwise and a large cistern empties through this pipe about four times each hour and washes excreta down the sloping corrugated floor and then to a drain. These pens need thorough cleaning only one or two times per year. As water is supplied for drinking by a trough at the rear of the pens and as feed is delivered with a scoop from a hopper on wheels, very little labour is required for maintenance of sheep in these pens.

If pens are raised above the animal room floor, trolleys for moving sheep should have their floors at a similar height to the pen floor height. Such trolleys can be used for holding sheep during experiments and so having the sheep at some distance from the floor is not a disadvantage as it obviates bending. However a ramp will be needed to walk sheep from the floor level into their pens as lifting heavy sheep places a strain on the lifter's back (see Fig. 15).

Fig. 15 *A sheep in a trolley designed for transport or holding for experimentation.* One side has been removed. The sheep is standing on a wire mesh below which is a collector for excreta.

Wire mesh is often welded to light metal frames of galvanized iron piping to form panels and these are allowed to rest on suitable supports to form pen floors. Such panels then can be removed for thorough cleaning underneath. Wire mesh should be of heavy gauge (4–5 mm diameter) to avoid its bending with holes no greater than 2·5 cm^2 so as to prevent feet being caught. If the holes are less than about 1·5 cm, faecal pellets will not readily pass through. Such mesh is not suitable for young lambs and if occasionally a ewe lambs in a pen, the mesh should be covered with matting. An alternative metal floor can be made from "expanded" metal (a heavy gauge metal sheet from which holes have been punched at regular intervals). If wooden slats are to be used, they should preferably be of hard wood, about 2 × 5 cm nailed to heavier timbers so that gaps of 1·5 cm are left between slats. Wooden flooring is also best made in removable sections.

Pathological changes in joints in sheep kept on slatted floors for 12 months have been reported in one paper.[26] These could be due to inactivity of the joints.

C. *Sides*

Pen sides are often made from wire or metal mesh welded or otherwise fastened to metal or wooden frames. They can be designed to be removable so that single pens can be converted into double or group pens. Commercially manufactured metal gates may be suitable. Wooden rails are less suitable for sides of pens unless placed closely together because sheep may slowly chew through wooden rails (this can be one form of stereotype behaviour (p. 126)). A satisfactory height for pen sides is 90 cm, although quiet sheep can be contained by a lower height.

D. *Water*

As with other laboratory animals, an ample supply of fresh drinking water must be provided. In many animal houses, this is often in a bucket held in a metal hoop attached to the side of the pen. Such a bucket needs to be removed, emptied and refilled every day or two. A long trough (about 12 cm wide and 9 cm deep) attached to the wall at the back of the pen and fed from a cistern is an alternative. Sheep sometimes defaecate into a water trough and hence a long trough should slope slightly to one end where there should be a large hole closed by a bung. It is sufficient then to clean the trough by removing the bung and replacing it a few minutes later when the water and any faeces have drained. A third method of supplying water is to use commercial large animal "waterers". These have a valve which is activated by pressure from the sheep's nose when it drinks.

Waterers, troughs or buckets can be at any height up to about 50 cm.

E. *Feed*

When sheep are kept on a solid floor, hay can be placed on the floor but much is wasted. There is less waste if the hay is placed in hay nets hung from a side of the pen.

If chaff, grain or pelleted foods are given, these may be placed in feed troughs on the floor or attached to the pen side or in 6–10 litre plastic buckets attached by hoops to the pen side.

G. *Temperature*

The thermoneutral zone for sheep depends on the wool covering (p. 87). Most laboratory animal houses are at a temperature comfortable for a recently-shorn sheep but hot for one with long wool. The latter type of sheep may pant for long periods. If possible, sheep with long wool should be accommodated in a cooler part of the animal house. Winter temperatures as low as 5–10°C are comfortable for sheep with reasonable fleeces.

6.2 Metabolism Crates

For many experiments, sheep are placed in metabolism crates. These crates allow the animal to stand or sit but not to turn around and hence are of great value when leads or infusion lines are attached. In some experiments, sheep are kept in metabolism crates for several weeks and in the author's department 50 sheep were kept in crates for 8 weeks in one experiment without apparent discomfort or harm. On the other hand, some people believe that sheep should not be so restrained for longer than 1 or 2 days. It is difficult to provide an answer as to whether or not long-term confinement in metabolism crates involves cruelty. Regulations in some countries allow for traditional laboratory animals to be able to turn around but include sheep with farm animals for which any such requirement is omitted. Possibly the reason for this is that special stud animals are often restrained so that they cannot turn around. Observations of sheep in metabolism crates indicate that, provided that they have been accustomed to the laboratory before introduction to the crates, they do not resent the greater confinement and that when removed periodically for weighing they are content to return to their crates.

Several designs for metabolism crates have been published.[6,7,15,23,30] Crates normally have a floor of metal or expanded mesh at a height of about 75 cm above ground level. The sides and rear have solid panels to deflect excreta to the crate floor but which can be removed for experiments. Excreta which falls through the floor is directed into a container by a funnel-like structure. When needed, the urine can be separated from faeces by a separator. At the front of the crate is a feeder and a water container (Fig. 16).

A typical size for a metabolism crate is 50 cm wide by 100 cm long with sides 60 cm high. These sizes might need to be increased sightly for very large sheep. A small sheep may need to be prevented from turning around by putting a dog collar around its neck and fastening the chain to the front of the crate.

Crates are probably most easily made from galvanized water pipe. This may be welded or joined with threaded elbows, T pieces and other plumbing fittings. Satisfactory crates have been made from lengths of pipe held together with Kee Klamps or from slotted steel metal and nuts and bolts.[3] The most common type used in Australia[30] is made from welded 20 and 25 mm pipe. Sides and excreta collectors are galvanized iron, polyvinylchloride plastic, fibre glass or stainless steel. Stainless steel is the most expensive material but is most easily cleaned and should be considered if many experiments involving radioactive isotopes are planned.

During experiments, metabolism crates make good devices for restraint. Other means have been used and Fig. 17 shows "an unsedated sheep on a standard supermarket shopping cart in a plastic semi-cylinder containing perforations for the extremities".[19] The device as shown could be improved by allowing the sheep's legs to touch the ground. Several workers have used a canvas sling with holes for the legs, the sling being fitted to a frame on wheels.

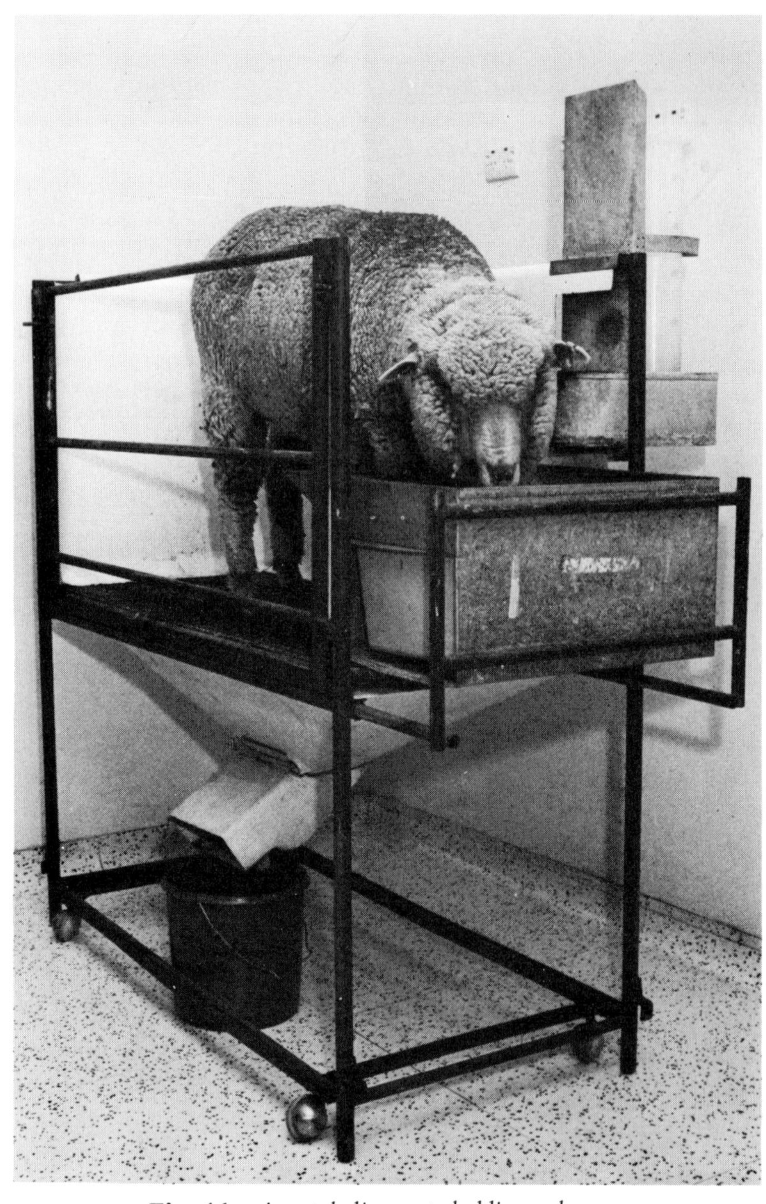

Fig. 16 *A metabolism crate holding a sheep.*
When not used for experimentation a fibre glass panel fits over the exposed side.

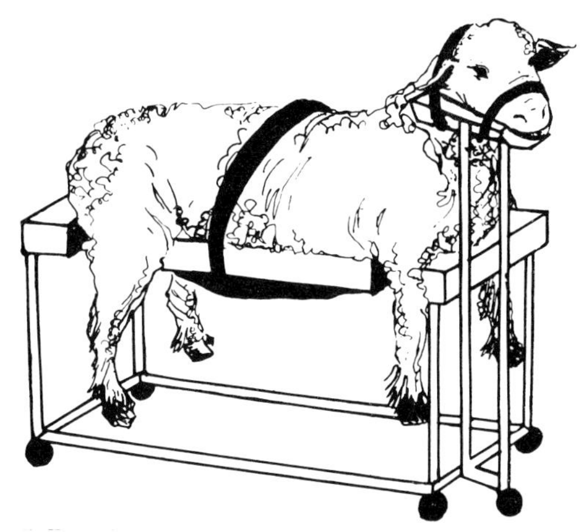

Fig. 17 *A sheep restrained for experimentation.*
It would be preferable to allow the sheep to rest its feet on a solid support.
[From Landa *et al.* by permission of the American Physiological Society.]

6.3 Identification

Several methods are used for identification of sheep. Permanent identification is provided by fire branding of numbers on the horns of horned sheep or by tattooing numbers on the skin of the inside of the lip or ear or under the tail. Tattooing pliers and sets of numbers are available commercially. The most commonly used means of identification is provided by either notching the ear according to a plan which shows units, tens and hundreds (Fig. 18) or by making a hole in the

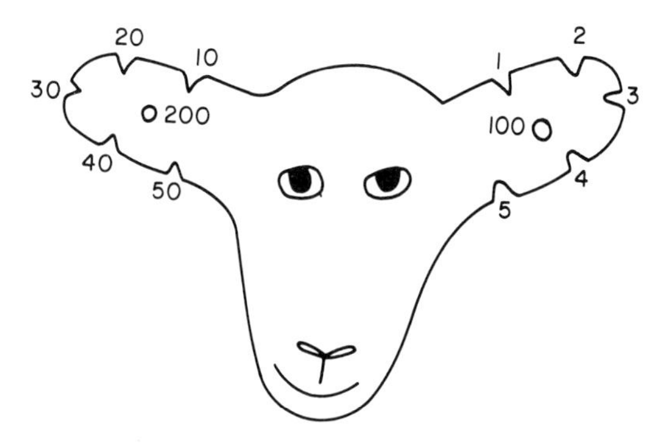

Fig. 18 *A code for identifying sheep by ear marks.*

pinna of the ear and inserting a numbered plastic or metal tag. Plastic tags are reusable and are made in several colours and so different colours can be used for different experiments. Occasionally an ear can become damaged such that a notch is lost or a tag pulls out. This is more likely to happen to sheep which are run in fields.

For new born lambs, tags designed for wings of poultry can be inserted into the ear.

When several rams are kept and some have been vasectomized, the tip of one or both ears is often cut off at the end of the vasectomy operation so that the ram can easily be identified as being infertile.

Temporary identification is best provided by colours sprayed or painted on the back or sides of the sheep. If the wool is to be sold, such colours should be scourable so that they do not spoil fabrics made from the wool. Spray aerosols with blue, red, green or purple are available. Marks last indefinitely on sheep in the laboratory but need to be renewed every 2 or 3 months on sheep in the field as sun and rain causes them to fade.

6.4 Routine Maintenance

Two types of clippers are needed for sheep as wool will need to be removed from areas for a variety of reasons including preparation for

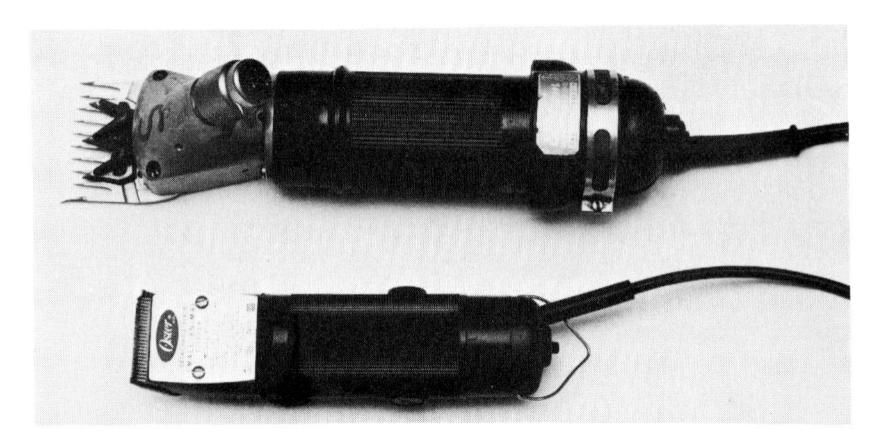

Fig. 19 *Two types of clippers used for removing wool from sheep.* The top set are used for removing long wool while the bottom pair are for fine wool.

surgery and for blood sampling. Hand blade shears can be used but electric clippers fitted with combs and cutters as used for commercial shearing of sheep make the task much easier. Once the longer wool has been removed with these coarse clippers (Fig. 19), electric clippers of the type used by barbers are used to remove the remaining few mm of wool. A number 40 head is best for fine wools which are the most difficult type to cut. Cutting surfaces will stay sharper if they are oiled each time before use and if they are not left running for a long period when not in contact with the wool. Lanoline in the wool lubricates them while cutting.

Occasionally a depilatory product as used by women is used. A depilatory mixture can be made up and applied for removal of wool from small areas. A suitable aqueous mixture is sodium sulphide (2·8%), sodium hydroxide (0·5%) and a wetting agent (2%). It must be stored in an air tight container. Depilatory agents act by dissolving the wool fibres below the skin surface where the wool proteins are susceptible to their action.[10]

Sheep require little routine maintenance in the animal house. Every 2 or 3 months, wool may need to be clipped from around the eyes if it impedes vision and from around the pizzle of rams and wethers and the vulva of ewes if it is regularly soiled with urine. Coarse clipping is all that is needed.

There is no wear of the horn of the hoof in the laboratory as is normal in the field and so the hoof continues to extend in length. If allowed to grow too long, the elongated horn distorts the hoof and causes discomfort. Hence each animal should be sat on its rump to have its feet trimmed at intervals of 4–6 weeks. Tree-pruning shears can be used. The dead horn is trimmed back to a natural shape by cutting near the quick. This is easy to do provided that the hoof is not overgrown but is difficult with neglected feet.

The only other routine maintenance required is regular clipping of wool from around cannulae and any surgical modifications such as carotid loops.

A. *Shearing*

A problem that is unique to the sheep as a laboratory animal is provided by the growth of the fleece. With commercial management, it is usual to shear sheep once each year and this is often done in spring so that lambing is made easier for ewes, so that they are cooler in summer and so that a reasonable fleece can grow before the next winter. In the laboratory, provided that animals are not infected with lice or ked and provided that the animal house and laboratory temperatures are not too high (so that sheep do not pant for long periods), there is no necessity for regular shearing. If the services of a professional shearer are readily available, it would be best to have sheep shorn once each year and a good time would be just before members of the laboratory take their annual vacations so that there is an interval for animals to recover from the effects of a sudden loss of insulation. In the absence of a shearer, it does not harm to let the wool grow for 2 years or longer.

The income from the sale of wool is unlikely to be of importance to a laboratory unless it has many sheep. A professional shearer (who may be difficult to find in a large city distant from sheep raising areas) might do the shearing in return for the wool and so save the laboratory staff the problem of disposing of it. Before shearing, areas surrounding surgical modifications such as cannulae, catheters and carotid loops should be clipped with fine clippers by the animal house staff so that these features can easily be seen and thus avoided by the shearer.

If a professional shearer is not available and a sheep has to be shorn, there are two options—the animal house staff may shear it or chemical defleecing may be used. Shearing is a difficult art to learn for it involves being able to restrain the sheep in several positions while wielding the shears. An unorthodox solution is to anaesthetize the sheep lightly, lie it on a table on its side and then remove the wool with clippers. The start is made along the mid-back line and the wool is removed from one side with "blows" made parallel to the back line until the legs are reached. Then the wool is removed from the legs, neck, head and abdomen before the sheep is turned over. The whole process will take 10 to 15 min. Coarse electric clippers are used and the blades should be kept against and parallel to the skin to avoid the necessity for making second cuts. The clippers should be allowed to advance through the wool at their own pace and should not be forced.

The alternative is chemical defleecing. Much research has been done on this subject in Australia and the United States with the aim of reducing shearing costs. The disadvantages of chemical defleecing in commercial flocks are that some wool may be lost in the paddock and that the recently defleeced sheep has no protection from sunburn.

Cyclophosphamide has been the drug most commonly used, at a dose rate of 20 to 30 mg/kg given by mouth. It is thought to stop mitosis for only a few hours and to have no apparent toxic effect. However, there may be a breed difference in susceptibility with Merinos being more susceptible to its effects and having wool growth stopped for some 2 weeks.[27] With chemical defleecing, the narrowing of or the break in the wool fibre allows the fleece to be lifted from the sheep some 2 to 3 weeks after dosing without causing discomfort.

6.5 Restraint

A good shearer will always hold a sheep so that it believes that its head is restrained and this is the secret of restraint of sheep. For a standing sheep, it is best for the holder to stand with his leg next to the neck with one hand around the top of the neck holding the head up slightly. The animal may attempt to move backwards if not well trained and should be allowed to back against a wall or into a corner. This method of holding provides good exposure for a second person to take a sample of blood from the jugular vein. Blood sampling can be done alone by one person in which case he stands astride the sheep's shoulders with the neck bent slightly around one leg. The elbow on the same side as that leg is pressed into the top of the neck for restraint of the head while that hand tenses the skin and occludes the vessel for the sample to be taken (Fig. 20).

Another method for restraint is for the animal to be sat on its rump and held between the holder's legs. The head may be held with one hand or the neck may be bent over (Fig. 21). Ruckebusch[28] has suggested that sheep in this position are in a trance-like state. To get the animal into this position, it may be grasped from behind with hands placed behind the forelegs, lifted and sat on its rump. This can be done with ease for small sheep but it places a considerable strain on the lifter's back. A preferable way is to stand alongside the sheep with one hand on the rump and the other around the jaw. A sudden push downwards with the hand on the rump and a simultaneous twist of the neck away from the holder will cause the rear end to sink to the ground. It is simple then to pull the sheep onto its rump.

It may be desired to have the sheep lying on the ground. To make it adopt this position, stand by the left shoulder and hold the head with

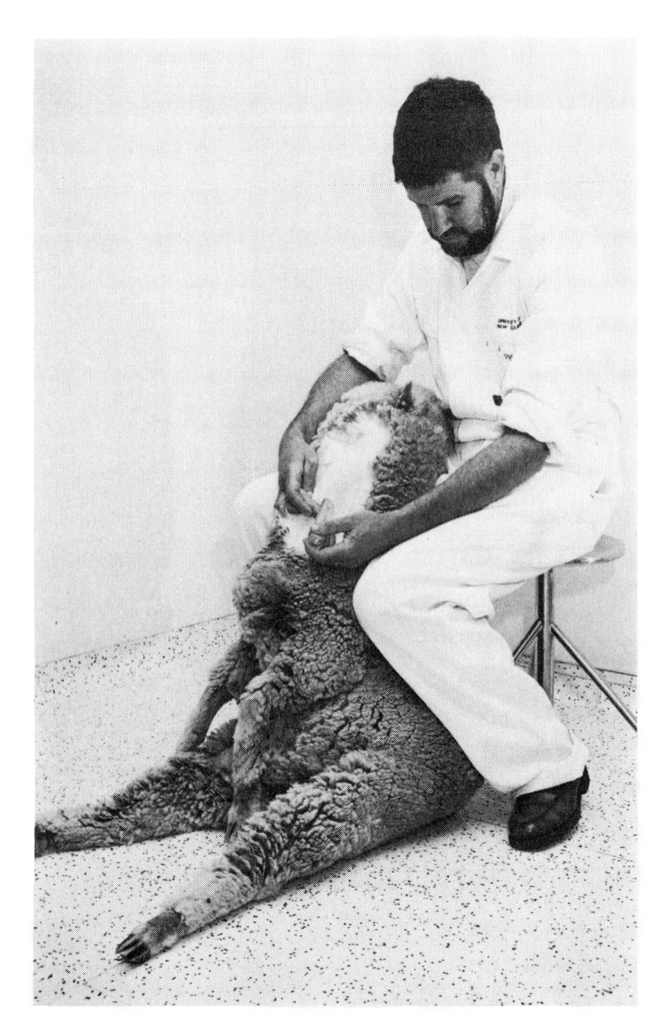

Fig. 20 *A method for holding a sheep for blood sampling.*
The second finger of the right hand is occluding the jugular vein while the left hand has directed a hypodermic needle into the vein.

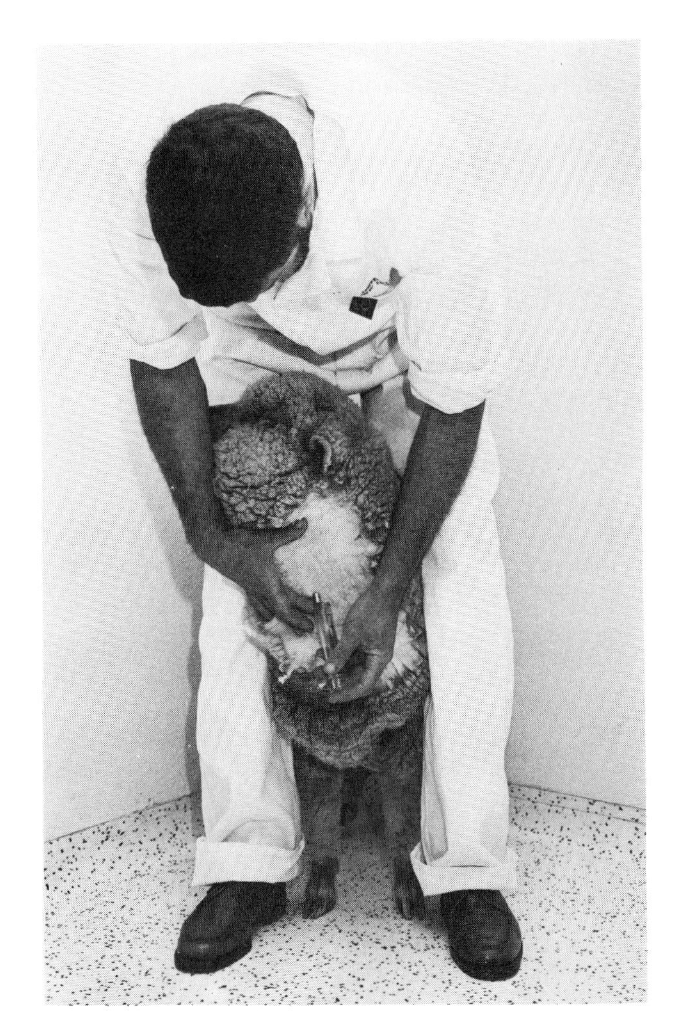

Fig. 21 *Another method of holding a sheep for blood sampling.*

the left hand while passing the other hand behind the near foreleg to grasp the offside foreleg. A pull on the offside foreleg while simultaneously pushing against the near shoulder will cause the sheep to lose balance and slide to the ground. Once it is on the ground, it will struggle least if the head is held and the feet are kept off the ground.

Sheep farmers who wish to transport a sheep from one place to another (possibly on the back seat of a car) will tie the legs together. One foreleg is tied to the opposite hindleg with twine or light rope, a third leg is added and then the fourth leg. If the head is covered with a cloth, the sheep will lie without struggling for a long period but to avoid any possible risk of damage to the circulation to the lower leg, a maximum time of 30 min should not be exceeded.

When catching a sheep in a yard, people unaccustomed to sheep will attempt to restrain it by grabbing at the wool. This is equivalent to and inflicts similar discomfort to grasping a person's hair. A group of sheep in a yard or a large pen will circle around the yard or pen away from a person trying to catch one. The circling should be allowed to continue until the wanted sheep is at the rear of the group. The catcher should then advance rapidly and grab one of the hind legs of that sheep. This will stop the sheep but it may attempt to shake the leg free and so restraint should quickly be transferred to the head.

6.6 Feeding

Because of the importance of sheep in meat and wool production, there has been much research done on their nutritional requirements. The presence of the ruminant-type stomach complex means in general that these requirements are not nearly as complex as the requirements of monogastric animals. To illustrate this point, sheep can live on a diet consisting of wheat straw treated with alkali to remove the protein, urea as a nitrogen source for protein synthesis and a supplement of a small amount of some of the branched chain amino acids and minerals. The treated straw can be replaced by molassus which is largely a thick solution of sucrose. On a straw based diet, if urea is given at a rate of 2% (equivalent to 13% crude protein), sheep do as well as if they were eating natural forage with 8 to 11% crude protein.[16]

In effect, ruminants can do well with cellulose, starch or sugar as an energy source and a nitrogen source which need not be protein. The energy source may be calculated in terms of metabolizable energy (ME), digestible energy (*DE*), starch equivalents (*SE*) or total digestible nutrients (*TDN*) and the protein source as either crude protein (nitrogen content × 6·25) or digestible protein.

Daily requirements[24] are 0·50 mJ of *DE*; 0·41 mJ of *ME* or 0·0123 kg of *TDN*/$kg^{0.75}$ liveweight for energy each day and 4·8 g of digestible protein/mJ of *DE*. For a sheep weighing 50 kg, the calculated requirements are 1 kg of dry matter per day containing 0·25 kg of *TDN* 10·07 mJ of *DE* or 8·24 mJ of *ME* and 89 g of total (crude) protein, or 48 g of digestible protein. Other requirements are 3 g of calcium, 2·8 g of phosphorus, 1·9 mg of carotine or 1275 i.u. of vitamin A and 278 i.u. of vitamin D. These requirements are for grazing animals gaining 10 g weight/day and hence are very liberal for a laboratory animal which does no walking and which probably is not intended to gain weight. For laboratory sheep, the above requirements are possibly 40 to 70% in excess of needs.[2]

It should be noted that sheep do not convert carotine to vitamin A in the same ratio as rats. It is suggested that 1 mg of feed carotines are the equivalent of 400–700 i.u. of vitamin A rather than 1667 for the rat.[24]

An alternative British system[21] is based on the formula:

$$ME = 1{\cdot}2 + 0{\cdot}13\,W$$

for growing and fattening sheep where *ME* is the maintenance requirement of metabolizable energy in megajoules (mJ) and *W* is the weight of the animal in kg. The following are recommended allowances with a correction for sheep being housed indoors:

Bodyweight	20	30	40	50	kg
ME requirement	3·8	5·1	6·4	7·7	mJ/day

Table 40 shows values for *DE*, *ME*, *TDN* and total (crude) protein for some typical foodstuffs fed to laboratory sheep. Straws have the lowest energy content and grains the highest, about twice that of the straws. With the exception of the straws and possibly wheaten hay and maize grain, the digestible protein levels are similar to or greater than the 8·6 g/mJ of *DE* recommended by the National Academy of Sciences. Thus except for the straws which require supplementation with a nitrogen source, it would be difficult to produce serious protein deficiency in laboratory sheep by feeding common feedstuffs. Also with the exception of the straws, it would be difficult not to supply sufficient energy by giving the recommended 1 kg of feed/day for a 50 kg sheep. The problem instead tends to be one of giving sufficient bulk to satisfy appetite without animals in the laboratory gaining excessive weight. It is only when animals are heavily pregnant, lactating or losing large amounts of protein because of frequent sampling or open lymph duct cannulae that supplementing roughage feeds with concentrates needs to be considered.

Table 40 Energy and protein contents of some feedstuffs (adapted from [24]).

Feedstuff	*DE* (mJ/kg)	*ME* (mJ/kg)	Crude protein (g/kg)	Digestible protein (g/kg)	Ratio of digestible protein to *DE*
Lucerne hay	10·0	8·4	14–20	9–15	9–15
Fescue hay	10·3	8·4	9·5	5·1	4·9
Oaten hay	9·9	8·1	9·2	5·1	5·1
Oaten straw	7·9	6·5	4·4	0·4	0·5
Wheat hay	9·7	8·0	7·5	4·0	5·0
Wheat straw	7·0	5·7	3·6	1·5	2·1
Barley grain	15·8	13·0	13·0	10·3	6·5
Maize (corn)	17·3	13·8	10·0	7·8	4·5
Oaten grain	13·8	11·3	13·2	10·3	7·4
Wheat grain	16–22	13–29	12–16	9–12	5·5–7·4

In practice, sheep are given hay, chaff, grains, "mixtures" or pelleted foods. Chaff is hay cut into lengths of 1 to 2 cm. Grains are preferably mixed (i.e. contain several types of grains) and crushed so that the individual grains are broken and consequently have a higher digestibility. Mixtures consist of a variety of foodstuffs and may contain one or more types of grain, probably some chaff or dried and milled grass, possibly some local material such as peanut hulls or dried brewers grain, and a small amount of selected minerals. Mixtures are usually made in bulk by an animal feed manufacturer. To make pellets, such mixtures are bound together with a little molassus, moistened, extruded under pressure through a 5–10 mm hole and broken into pieces 1–2 cm long.

Storage of food in an animal house can cause a space problem. Hay is very bulky and requires the greatest storage space. Chaff is slightly less bulky and is more easily handled. Hay and chaff are often used at agricultural research institutes where they are produced on farms attached to the institutes. For other places, mixtures or pellets are preferable as they are less bulky, are associated with less dust and can be stored in hoppers which allow gravity feed into a bucket or small trolley for feeding out.

Although nutritive requirements can be calculated from sources of data such as those cited above, it is rare that calculations are done for laboratory sheep. Many scientific papers state that the sheep were fed *ad libitum* which usually means that they were fed a supramaintenance diet and were gaining weight. Such experiments are usually of a short-term nature and for a long-term experiment, it is preferable that animals keep a relatively constant weight. To do this, initially a dipper containing 0·75–1·0 kg of feed is given each day and animals are weighed every 2–4 weeks. If it is found that they are increasing in weight (allowance being made for wool growth: 0·25–0·5 kg per month depending on breed), a larger or smaller amount should then be given. Sheep will usually eat all of a maintenance ration in approximately 2 h and feed containers should be empty next day.

Lucerne hay or chaff contains adequate carotine and vitamin D and commercial mixtures or pellets probably will have had these vitamins added. Some diets fed for a long period may require supplementation although Tucker *et al.*[31] were unable to produce symptoms of vitamin A deficiency in sheep fed a deficient diet for 30 months. Supplementation can be done regularly but it is simpler to give oral doses of each of these vitamins every 3–6 months if the diet may be deficient as both are stored in the liver. Poisoning with vitamin D has been reported in sheep given 5 000 000 i.u./animal.[5]

When the diet consists largely of grains, the calcium level should be considered as grains have the calcium to phosphorus ratio greatly imbalanced towards the latter element. High phosphorus intakes relative to calcium are likely to lead to urinary calculi and possibly to osteoporesis. To prevent this, a little powdered chalk may be sprinkled on the food. Urinary calculi have been reported as being caused by feeding pellets with low calcium but high magnesium content and intended for supplementing sheep grazing pastures low in magnesium.[18] Another dietary problem with commercial foodstuffs that has been reported was with pellets that were intended for other species, contained large amounts of copper and caused copper poisoning (p. 206).

For someone who wishes to start using sheep as laboratory animals, it is recommended that they contact a supplier of animal foodstuffs for a regular supply of sheep pellets or a mixture suitable for sheep. To avoid compromising experiments by changes in feeds, the feed chosen should be of a relatively standard composition and a long-term supply should be assured. When feeding first starts, the requirements for maintenance should be determined by trial and error. Then occasional weighing of animals could be done to check that the ration is given at a maintenance level. Lactating ewes, heavily pregnant ewes and sheep which are being used for intensive sampling may then be given supplements as necessary.

Although pellets are widely favoured as feed for laboratory sheep, they have some features that induce changes in the digestive physiology. The main change is that the roughage in pellets is in a fine state due to prior milling. The fine state of the fibre reduces the desire for rumination and and sheep fed pellets will ruminate for shorter periods. Also the food passes through the reticulo-omasal orifice more quickly with the result that the size of and digestion in the rumen are reduced and the size of and digestion in the large intestine are increased. These changes are unlikely to compromise experiments other than in digestive physiology. Surgeons find that surgery in the abdominal cavity of pellet-fed sheep is easier due to the smaller size of the rumen.

Occasionally sheep brought freshly from the field and which are not accustomed to supplementary feeding may not immediately start to eat in the laboratory. Nothing should be done for 2 or 3 days. Also sheep recovering from intestinal surgery may be reluctant to start eating due to ileus. The appetite of sheep can be stimulated in various ways. They may respond to a small amount of salt or dried molassus sprinkled on the food or if available some fresh grass. Sometimes the presence in the same pen of a second sheep which readily eats laboratory food will tempt a reluctant sheep to start eating.

When the diet is changed substantially, it takes in the order of 10 days to complete the major adjustments in the rumen microbial population[32] although full stabilization of numbers may take 3 weeks.[14] Changes in the level at which a ration is given usually have little effect on the microbial flora.[32]

Labour is the most expensive component in the care of laboratory animals. For many experiments, it may be sufficient to give sheep feed on each weekday morning and then give double rations late on Friday afternoon rather than requiring someone to come to the laboratory specially during the weekend to feed.

Sheep fed once daily will finish their daily ration within a few hours and as a result are "starved" for the remainder of the 24 h. This creates a situation different to that of the grazing animal and induces a marked diurnal rhythm in gut function which can impose a diurnal rhythm on other body functions.[33] Such a rhythm can be eliminated by the use of a "continuous" feeding system. The simplest type of continuous feeder consists of a continuous wide rubber belt mounted above a metabolism crate so that feed falls off the front end of the belt by the belt's slow rotation into a chute which directs the feed into the animal's feed container. The belt is driven by an electric motor through a gearing system so that the belt is advanced by 1 or 1·5 m each 24 h.[4,12,22] Alternatively a timing mechanism may control the motor so that the belt advances for a short period each hour by a distance of 4·2 or 6·3 cm (1 or 1·5m/day). To use this system, the day's feed is spread evenly along the 1 or 1·5 m of the belt at the same time each day. Other designs of "continuous" feeders have 24 small bins, one of which is caused to tip into the chute each hour, or sometimes eight bins of which one tips each 3 h.[17] Such continuous feeding may not eliminate all cyclic fluctuations in rumen parameters.[8]

Although most foods are extensively digested in the rumen, there is a technique for passing some food components through the rumen undigested. Proteins in the rumen must be in solution to be digested (p. 70) and the technique relies on coating small particles of the food with a suitable protein such as casein and then treating the protein with formalin to render it insoluble in the rumen but digestible in the abomasum and small intestine. The technique has been used to supply special proteins to the intestines[10] and to enrich the depot fats with polyunsaturated fatty acids.[29]

A. *Orphan lambs*

While the lactating ewe provides the most convenient method for feeding lambs, lambs can be reared quite successfully by hand. Commercial dried milk preparations for young animals should be used for economy with large numbers but for a small number preparations intended for baby feeding are suitable. After the first few days, ordinary cows' milk can be substituted. Special teats for lambs which are much stiffer than baby teats can be obtained but baby teats may be used if the hole in the end is enlarged and the teat is held firmly on the spout of the bottle to prevent the vigorous sucking of a hungry lamb from pulling it off.

A lamb which is only a day or two old is easier to train to bottle feeding than an older lamb. A healthy lamb of this age will seek avidly for the teat after the first feed or two and the temptation is to overfeed. Overfeeding will lead to digestive troubles. For the first few days, the milk should be given in four or five feeds each day in an amount that will cause the lamb's flanks still to look slightly hollow. After the first few days, the number of feeds can be reduced to 3/day (9 a.m. and 1 and 5 p.m. suit most laboratory schedules). Grass or good hay and a commercial concentrate mixture for raising young ruminants should be offered as early as it will be eaten (3–5 days). Weaning may be contemplated after 3 or 4 weeks if the lamb is eating well.

6.7 Management of Experimental Sheep in Paddocks

In many places, reserve sheep for experimentation are kept under field conditions rather than in the animal house. Also for some research projects, especially those relating to agricultural research, sheep are kept in paddocks. Often they are part of a general farm flock which fits into the management procedures used by the farm. It may however happen that a research institute wishes to keep sheep under field conditions but not in conjunction with a farm flock. To do this, standard texts on sheep raising should be consulted.[1,11,25] The following are some hints of relevance to experimental rather than production management. It is assumed that a well fenced field with an ample water supply is available.

The stocking rate will depend on the type of pasture that is grown and on factors such as average rainfall and temperature. It may be necessary to provide supplementation if many sheep are to be run on a small area or as seasonal conditions dictate. Either hay or the same feedstuffs that are fed in the laboratory can be used. Some shelter may be necessary if heavy snow falls are likely but local advice should be obtained on such matters.

Usually field sheep require little attention. They will need to be yarded occasionally and hence a yard in a corner of a paddock with a

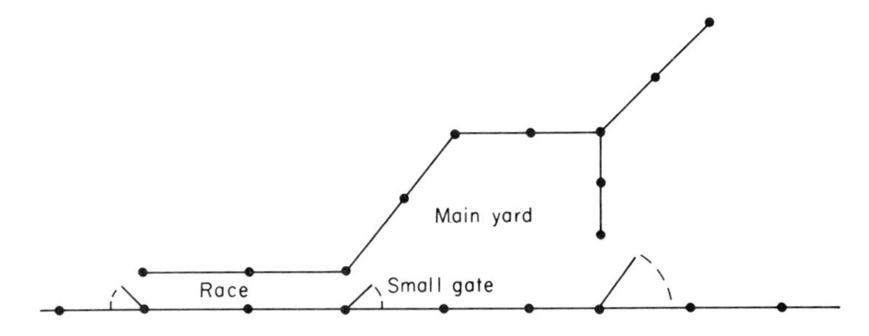

Fig. 22 *A simple yard for a small flock of experimental sheep.* It is built along an existing strong fence.

race leading off it is essential (Fig. 22). The periodic attention that they do need includes shearing once per year, regular drenching for internal parasites and control of blow fly strike. Internal parasites are primarily a problem of young animals, of animals in a slightly precarious nutritional state or where there is overstocking. Sheep, like other animals will develop a reasonable degree of resistance to internal parasites which is one of the reasons why parasites are less of a problem in older animals. Usually two or three times each year there are conditions which are most favourable to infection of sheep with immature stages of internal parasites and local advice should be sought as to the optimum times to treat sheep with anthelminthics in order to keep infestation of the pasture to a tolerable level. Sheep which are not "doing" well should be suspected of having a high worm infestation and faecal samples should be taken from those and also from a few others in the flock. Soiling of the perineum with faeces is one sign of an excessive worm burden. If a heavy infestation is found, a change in drenching routine should be adopted.

Fly strike can be a problem, especially when wool is long and the weather is damp. For treatment of fly strike, see p. 202. Prevention consists of clipping areas of wool likely to be soiled such as around the vulva or pizzle and spraying the wool with an insecticide delivered with a high pressure spray. If one of a flock is struck, the likelihood is that others may soon also become struck and so preventative measures should be considered.

Another problem is caused by domestic dogs running free. Dogs enjoy chasing sheep and often will band together in packs to hunt sheep into a corner. Then some may walk on top of the flock and bite pieces of flesh from their backs. Those bitten are likely to die from clostridial infection but some not severely damaged may be saved if given antibiotics. The presence of numerous domestic dogs in an area may make the maintenance of sheep there impossible unless a dog-proof fence is erected.

6.8 Disposal

Sheep may need to be "put down" at the end of an experiment. In an abbatoir, death is usually caused by swiftly severing the neck vessels dislocating the neck. This requires much practice to do properly and is not recommended. If a "captive bolt" type of humane killer is available, this should be applied to the forehead at the point of the intersection of lines drawn between the eyes and the contralateral ears. Alternately, animals may be lightly anaesthetized before a major artery such as the carotid is severed.

Disposal of carcases is usually by burial or incineration. A chest-type freezer is handy for storage of carcases before disposal. A woolled carcase may take a considerable time to cool in a freezer, especially as wool limits heat loss and as fermentation continues in the rumen for a long time after death. As a result, the abdomen may distend with gas. Prior removal of abdominal viscera is recommended.

References

1. Belchner, H. G. (1976). *Sheep Management and Diseases*. Angus and Robertson; London.
2. Blaxter, K. L. (1962). (See reference 44 in Chapter 3.)
3. Church, D. C. (1969). (See reference 79 in Chapter 3.)
4. Clapperton, J. L., Thomas, P. C., Stokes, M. R. and Henderson, J. M. (1974). A compact machine for feeding sheep up to twelve times daily. *Br. J. Nutr.*, **31**: 271–272.
5. Clegg, F. G. and Hollands, J. G. (1976). Cervical scoliosis and kidney lesions in sheep following dosage with vitamin D. *Vet. Rec.*, **98**: 144–146.
6. Dick, A. T. and Mules, M. W. (1954). Equipment for the clean collection of twenty four hour samples of urine and faeces from sheep. *Aust. J. agric. Res.*, **5**: 345–347.
7. Duthie, I. F. (1959). A sheep metabolic cage for mineral balance and radio-isotope experiments. *Lab. Pract.*, **8**: 408–412.

8. Faichney, G. F. and Griffiths, D. A. (1978). Behaviour of solute and particle markers in the stomach of sheep given a concentrate diet. *Br. J. Nutr.*, **40**: 71–78.
9. Ferguson, K. A. (1975). The protection of dietary proteins and amino acids against microbial fermentation in the rumen, pp. 448–464. (In reference 20.)
10. Ferguson, L., Laws, L. and Hopkins, P. (1976). Chemical crutching of sheep. *Wool Tech. Sheep Breeding*, **24**: **No. 4**, 23–25.
11. Goodwin, D. H. (1971). *The Production and Management of Sheep*, Hutchinson Educational; London.
12. Gray, F. V., Weller, R. A., Pilgrim, A. F. and Jones, G. B. (1967). Rates of production of volatile fatty acids in the rumen: V. Evaluation of fodders in terms of volatile fatty acid produced in the rumen of the sheep. *Aust. J. agric. Res.*, **18**: 625–634.
13. Greenhalgh, J. F. D. and Reid, G. W. (1974). Long- and short-term effects on intake of pelleting a roughage for sheep. *Anim. Prod.*, **19**: 77–86.
14. Grubb, J. A. and Dehority, B. A. (1975). Effects of an abrupt change in ration from all roughage to high concentrate upon rumen microbial numbers in sheep. *Appl. Microbiol.*, **30**: 404–412.
15. Harrison, F. A. (1974). The Babraham metabolism cage for sheep. *J. Physiol.*, **242**: 20P–22P.
16. Hogan, J. P. and Weston, R. H. (1971). The utilization of alkali-treated straw by sheep. *Aust. J. agric. Res.*, **22**: 951–962.
17. Jansen, T. H., Voster, P. W. and Basson, W. D. (1973). An electrically operated automatic feeder for sheep. *S. Afr. J. anim. Sci.*, **3**: 101.
18. Jones, J. O. and Dawson, P. (1976). A urological syndrome in young lambs. *Vet. Rec.*, **99**: 337–338.
19. Kung, M. *et al.*, (1978). (See reference 225 in Chapter 3.)
20. McDonald, I. W. and Warner, A. C. I. (1975), editors. *Digestion and Metabolism in the Ruminant*, University of New England Publishing Unit; Armidale, Australia.
21. Ministry of Agriculture, Fisheries and Food (1975). Technical Bulletin 33: *Energy Allowances and Feeding Systems for Ruminants.* London: Her Majesty's Stationery Office.
22. Minson, D. L. and Cowper, J. L. (1966). Diurnal variations in the excretion of faeces and urine by sheep fed once daily or at hourly intervals. *Br. J. Nutr.*, **20**: 757–764.
23. Mitchell, A. R. (1979). Robust and inexpensive metabolism cages for sheep. *Br. vet. J.*, **135**: 294–296.
24. *Nutrient Requirements of Domestic Animals: Nutrient Requirements of Sheep*, 5th ed. (1975). National Academy of Sciences; Washington, DC.
25. Owen, J. B. (1976). *Sheep Production*. Bailliere Tindall; London.
26. Preusse, C., Seffner, W., Bosemann, W. and Viehweger, E. (1980). Untesuchungen zur Gliedmassengesundheit von Merinofleischschafen bei Spaltenbodenaufstallung. *Monat. Veterinamed.*, **35**: 699–702.
27. Reis, P. J. and Chapman, R. E. (1974). Changes in wool growth and the

skin of Merino sheep following administration of cyclophosphamide. *Aust. J. agric. Res.*, **25**: 931–943.
28. Ruckebusch, Y. (1964). *Rev. Med. vet.*, **27**: 793–806.
29. Scott, T. W. and Cook, L. J. (1975). Effect of dietary fat on lipid metabolism in ruminants, pp. 510–523. (In reference 20.)
30. Till, A. R. and Downes, A. M. (1963). A metabolism cage for sheep. *Lab. Practice*, **12**: 1006–1009.
31. Tucker, R. E., Mitchell, G. E. and Little, C. O. (1967). Yellow pigments excreted by vitamin A-depleted sheep. *J. Nutr.*, **93**: 518–522.
32. Warner, A. C. I. (1962). Some factors influencing the rumen microbial population. *J. gen. Microbiol.*, **28**: 129–146.
33. Warner, A. C. I. (1966). Diurnal changes in the concentrations of micro-organisms in the rumens of sheep fed limited diets once daily. *J. gen. Microbiol.*, **45**: 213–235.
34. Williams, H. Ll. (1978). (See reference 48 in Chapter 4.)

Chapter 7

Sampling and Recording

For each species of experimental animal, there are techniques that facilitate sampling and obtaining of data. Some of these techniques are used for a range of experimental animals while others can be used only for a particular species. This chapter outlines some of the techniques that are used in various laboratories for sheep.

7.1 Blood Sampling

The easiest site by far for sampling of blood is the jugular vein. This vessel is separated from the skin by only a thin layer of fascia and is about 1 cm in diameter. Many research workers insert a catheter or cannula into the jugular vein for sampling but, if insertion is done several times inexpertly, the vein may become so damaged that future insertion is difficult. Often lengths of plastic tubing are inserted through a large bore needle to act as catheters. Difficulty in insertion can be caused by the use of a blunt needle or failure to clip the wool over the jugular vein properly. The alternative to inserting a catheter or cannula is to use a hypodermic needle and many samples can be taken in a day if the size of the needle is kept to 19 gauge or finer. With the method of restraint shown in Fig. 21, a 10 ml sample can be taken through a hypodermic needle in about 30 s from entering the sheep pen provided that the skin over the vein has been clipped previously.

If a catheter has been inserted into the jugular vein, the length of catheter in the vessel should be such that the tip does not lie at the entrance to the thoracic cavity. Autopsies done on a number of sheep with jugular catheters showed that if the tip was at this point, normal movements of the animal's neck caused the tip to rub against the vessel wall with the development of a massive thrombus which could grow to obstruct the lumen of the vessel. Lesions associated with the tip of the catheter when it was 5 cm before the entrance to the thorax were rare as were lesions when the tip was about 5 cm inside the thorax.[37]

Valvular endocarditis has been associated with the use of venous catheters in rabbits[67] and is seen occasionally in sheep that have had venous catheters. It is a criticism of many research workers, more of those in the agricultural than the medical research field, that sterility while using vascular catheters is rarely achieved. There are some reports in the literature of high mortality of sheep with catheters and these losses are without doubt due to catheter-induced infection.[3,58]

Venous blood can also be taken with a fine (19–21 gauge) needle from the cephalic vein if the sheep is sat on its rump (Fig. 20) and the wool over the vein is clipped. The saphenous or recurrent tarsal vein is not very useful as sheep will usually kick the back leg when the needle penetrates the skin. For a similar reason, the ear veins are not often used although they are suitable if the head is well held.

The alternatives for sampling of arterial blood are from a carotid loop (a surgical modification for sampling by which a carotid artery is enclosed in a skin tube—see [36]) or by direct puncture of the femoral artery. The latter artery can be palpated under a layer of muscle in the groin of a sheep sat on its rump but the groin of a sheep is dirty from accumulated secretions and should be well cleaned before sampling.

The volumes of blood that can be taken depend on the size of the sheep and its nutrition. In order to produce severe anaemia, Winter[93] took 360–540 ml daily for 9 days. This reduced the erythrocyte count, haemoglobin level and haematocrit by about 70%.

A. *Patency of vascular catheters*

In a previous review,[36] a point was made that many research workers have great difficulty in keeping intravascular catheters patent for long periods of time, not only in sheep but also in other experimental animals. A suggestion was made that it would be worthwhile testing various catheter materials to see which was the least thrombogenic. Subsequent to making that suggestion, a major research project has been started in which over 500 sheep have been used for determining the thrombogenicity of catheters. The project has particular relevance to catheters and cannulae used in humans as no testing of these in blood vessels is required before they reach the market. In the sheep tests, differences in thrombogenicity of over 13 fold have been found. As this has direct implications for animal experimentation, a brief summary of the more important results will be given. Further details may be found in [40,41,83].

The test model involves inserting tubings (catheters) into selected vessels using asceptic surgical techniques on sheep under general anaesthesia. Vessels which have been used are the facial vein (leading into the jugular vein and then into the anterior vena cava), the cephalic vein (into the anterior vena cava), the saphenous vein (into the posterior vena cava) and the femoral artery (into the aorta) of ewes and the posterior vena cava and aorta (via the femoral vessels) of new born lambs. Insertion is by cut-down and ends are closed and buried under the skin to avoid infection passing from the skin along the tubing. After a predetermined interval (often 9 days for ewes and 6 days for lambs), animals are heparinized, anaesthetized and killed by exsangination. Catheterized vessels are dissected and opened and the weights of thrombus along set lengths are measured. A square root transformation of data is needed before statistical analysis. The aortic catheters cause some renal infarcts and the surface areas of these have been measured in a few experiments.[38,41] The results show that the amount of thrombus increases with both the duration of catheterization and the diameter of the tubing. There were significant differences between materials with all polyethylene tubings tested being more thrombogenic than most silicone rubber tubings. However the differences between the materials were less marked than the differences within materials for silicone rubber, nylon, polyethylene and polyvinylchloride. The differences within materials appeared to be due mainly to differences in the roughness of the surface and for many types of catheters used in humans marked surface roughness was caused by particles of heavy metals which had been incorporated in the plastics to produce radio-opacity. In several, the radio-opaque material was present in a narrow strip which was rough while the remainder of the tubing's circumference was smooth.

Many of the formed tips of catheters sold for use in humans have rough areas when viewed in a scanning electron microscope and evidence was obtained that these tips could induce thrombus formation.[39]

The other finding of direct relevance to sheep experimentation is that heparinization of the animal was not very effective in reducing thrombus formation around catheters and that two platelet-inhibiting drugs, aspirin and dipyridamole, were of no use. For heparinization to be effective, the clotting time needed to be markedly increased and this required dosing in the order of 10 000 i.u. of heparin given subcutaneously every 12 h.[84]

7.2 Blood and Other Cells

Standard electronic particle counters can be used for sheep leucocytes,[50] platelets[86] and red blood cells (p. 41).

Sheep red cells do not aggregate spontaneously but can be induced to do so for separation of leucocytes by mixing five parts of blood with one part of a 1% solution of hydroxyethylcellulose which induces rouleaux formation.[47]

A method for obtaining sheep neutrophils is by the insertion of a small perforated plastic capsule into subcutaneous tissue.[34] Slow perfusion of this capsule will provide 25×10^6 neutrophils/h as well as some prostaglandin E_2. Almost pure lymphocytes can be obtained from lymph duct cannulae in efferent lymph ducts[80] and macrophages from the secretion of the involuting mammary gland of the ewe.[49]

Staining of blood cells may present difficulties as sheep blood and marrow cells are more difficult to stain than are human cells. May-Grunwald-Giemsa stain or Wright stain are best but they must be used at the pH optima.[92]

Measurement of microhaematocrit has been investigated and it was determined that centrifugation for 8 min at 10 000 *G* was optimal for the relatively small sheep red blood cells.[24]

Lymphocytes can also be prepared by removing erythrocytes from blood and then allowing the granulocytes and monocytes to stick to glass.[34] B lymphocytes (sIg +) can be detected by the direct antiglobulin rosetting reaction using anti Ig-coupled pig red cells.[8]

7.3 Respiration

Sheep tolerate well masks made to fit over the mouth and nostrils. Satisfactory masks have been made from a variety of materials including fibre glass and metal cans with pentrose tubing glued to the inside for a pneumatic seal or with a rubber diaphragm containing a large hole glued to the end.[85] One of the occasions when horns on a sheep are useful is when they are used to secure a mask onto the face; otherwise a halter must be used.

Endotracheal tubes are vigorously chewed if introduced into the mouth of conscious or semiconscious sheep, but are well tolerated if introduced through a nasal passage. The trachea of a conscious sheep can be cannulated over a flexible fibre-optic bronchoscope inserted intranasally and the sheep will show few signs of the interference apart from one or two coughs when the bronchoscope first passes the vocal chords.[48,87] A small tube can be passed into the separate bronchus to the right apical lobe of the lungs (p. 15).[27,74]

Permanent tracheostomies can also provide access to the respiratory system. If air is not allowed to pass over the upper respiratory tract, a sheep may have difficulty in keeping cool in a warm laboratory (p. 87).

A. *Respiratory rate*

Normal respiration can be counted easily by watching movements of the flank but panting occurs at too rapid a rate to be counted visually. A bellows type of pneumograph fixed around the posterior part of the thorax may be used. Preferable is a thermistor or thermocouple fixed over a nostril so that the sensing element is in the air stream. It may also be inserted via a guide tube previously inserted surgically through the skin and bone into the nasal cavity[42] or between two tracheal rings.[77]

7.4 Cardiovascular

A. *Heart rate and electrocardiogram*

The electrocardiogram provides the best means of determining heart rate. The orientation of the sheep's heart is different to that of man and so the standard lead postions for humans are not used with sheep. It has been suggested that the equivalent to the standard human leads can be obtained by placing (a) the right forearm lead on the head between the ears, (b) the left forearm lead on the sacrum, (c) the left leg lead in the mid-sternal region 2 cm anterior to the infrasternal angle and (d) a reference electrode on the left hind leg.[79]

The ECG trace is affected by movement of an animal and, although a sheep will stand still for long periods of time, it may move at a critical time. Probably the trace which is least affected by movements is provided by leads placed along the dorsal midline—between the shoulders, at about the level of the last rib, and in the region of the pelvis. These positions for leads even provide moderate traces from a sheep walking on a treadmill.

Other electrode lead positions that have been used are:

(a) implanted subperiosteally on the fifth rib at a point close to the apex of the heart and on the third and fourth thoracic vertebrae.[89]
(b) immediately cranial to the shoulder joint, on the abdominal wall immediately cranial to the left stifle joint, and midway between the right iliac crest and the right ischiatic tuberosity (for the reference electrode).[81]
(c) one high up on the animal's back on the left hand side and a second on the sternum.[63]

The best electrical contact is provided by needles or wires inserted through the skin. For short-term use, Michel clips may be used. Surface electrodes are rarely used but in one paper [63], they were used with a special electrode jelly and held in place with a harness.

B. *Blood pressure*

In the absence of an arterial catheter, measurement of blood pressure is not easy as there is no limb with a convenient large artery for manual palpation of the pulse. A normal size sphygmomanometer cuff has been used on the hind leg between the stifle joint and the tarsal region.[32,33,75] A pencil type of Doppler flow probe would be useful here for detecting the systolic end point. Most times when arterial pressure is to be measured, a carotid loop is surgically constructed before hand[36] and a pediatric-size sphygmomanometer cuff is used together with a small microphone.[21]

In few references where blood pressure values are cited is the vertical reference point given. A satisfactory reference point which corresponds approximately to the level of the atria is one third of the distance from the brisket (measured immediately behind the forelegs) to the top of the back.[97] As the carotid baroreceptors have a marked influence on blood pressure, the vertical position of the head should be noted when a reading is taken.

7.5 Faeces and Urine

Often the reason why sheep are kept in metabolism crates (p. 156) is to permit collection of faeces. When excreta fall through the mesh floor of the crate, they pass over a separator which allows urine to fall into one container while directing faecal pellets into a second container. Separators vary in their efficiency in providing "clean" separation but as sheep rarely defaecate and urinate simultaneously separation usually is adequate.

Although sheep defaecate several times each day, there are always some pellets retained in the rectum after each defaecation. Thus "grab" samples of faeces can be obtained by inserting a finger into the rectum. A few pellets can be withdrawn with the finger but the stimulus usually causes the animal to defaecate a minute or two later. If the sheep has been lying down in its pen, rectal stimulation may not even be necessary as most sheep will defaecate a minute or two after standing up.

A faeces bag attached to the perineal region by a harness is the easiest method for making a complete collection of faeces from a ram or a wether under grazing conditions and can also be used in an animal house. A faeces bag is not suitable for use with ewes although types of "metabolic harnesses" for ewes have been described.[2,4,22,43,57,90]

The bladder of the ewe can be catheterized with a standard human urethral catheter stiffened with a wire introducer. A vaginal speculum is needed as there is a suburethral diverticulum which lies ventral to the urethra and which can be entered. Practice makes urethral catheterization easy and once the technique has been mastered it can be done quickly (e.g. [88]). Some types of urethral catheters can irritate the urethra (especially types made from red rubber) and these should not be left in for more than a day or two. Other reports indicate that longer periods are feasible.[15]

Various types of urine-collection devices strapped over the abdomen of wethers and rams have been described.[2,10,11,22] For ewes, pentrose tubing or a rubber condom with a tube glued into the end which is attached with contact adhesive around the vulva works well for two or three days but then needs to be reglued.[71,85] Prior cleaning of the skin with isohexane is necessary for adhesion. When urine is to be collected from grazing sheep, the device of Chambers *et al.*[15] allows measurement of the total volume and collection of a sample of 1–10% of the output.

Sheep can be trained to urinate by touching them gently on the hind quarters[85] or by placing a hand over the muzzle to restrict breathing.[19,60]

7.6 Digestive Tract

A. *Rumen and jaw movements*

If a rumen fistula has been made, a rubber balloon can be inserted through the fistula on the end of a length of tubing and partially inflated. It may be possible to pass the balloon intranasally to the rumen, but the length of connecting tubing would dampen the pressure waves considerably and may interfere with rumination. Surgical creation of one or more small adhesions between portions of the rumen wall and the skin allows a mechanical recording of rumen contractions to be made.[72] A continuous record of jaw movements provides information on the pattern of eating and rumination. A good method is to hold a small balloon under the jaw with a head stall.[7,96] As the movements associated with rumination are regular, interrupted occasionally by a brief pause for swallowing and regurgitation of a new bolus, periods spent ruminating can easily be identified.

Rumen movements can also be detected from the "electrogastrogram" recorded from an electrode placed between the junction of the last ribs with the sternum and one placed to the left of the base of the tail. The record is of periodic displacement of a volume conductor when the reticulum contracts.[44]

B. *Rumen fluid*

A plastic tube can readily be inserted through a nostril and advanced into the oesophagus and then into the rumen for administration of solutions. This method may be easier for dosing a sheep than by mouth as sheep attempt to chew vigorously anything placed in the mouth. The maximum size of tube that can readily be passed through the nose is about 6 mm but this size is too small for sampling of rumen contents as the fibrous material in the rumen blocks the end of the tube. If it is necessary to obtain a sample of rumen fluid and the animal does not have a rumen fistula, then a plastic or rubber tube of about 1 cm diameter should be passed through the mouth into the oesophagus and rumen for fluid to be removed by suction. Difficulty due to blockage may occur even with this size of tube. The tube must be kept away from the molar and premolar teeth which have very sharp edges and which will rapidly sever the tube (or lacerate a finger). The sheep should be restrained by the holder straddling the shoulders and holding the head with a hand under the jaw. The tube is introduced through the gap between the incisor and premolar teeth and is held in the centre of the mouth with the index finger and thumb of the hand holding the jaw. The animal will attempt to move the tube laterally to between the teeth.

Regular samples of rumen fluid are more easily obtained through a rumen fistula. Several methods have been published for making such fistulae.[36] The possession of a rumen fistula has little effect on digestive parameters.[52] If rumen contents rather than rumen fluid are to be sampled, a large fistula through which a hand can be inserted should be made.[59] This will allow complete emptying of the rumen and reticulum if necessary and it will also allow tubes and other devices to be directed through the reticulo-oesophageal orifice or into the omasum and abomasum.

If a rumen fistula is not desired, a plastic tube for infusion and possibly one large enough for sampling can be inserted either directly through the rumen wall by means of a trocar and cannula[30] or through an adhesion surgically created between a portion of the rumen wall and the skin.[64] The same technique of creating an adhesion can be used also for other parts of the tract such as the abomasum, caecum and gall bladder.

C. *Digesta flows in the gut*

As much of the research on sheep involves the digestive system, techniques have been developed for facilitating measurement of flows of nutrients in the gut. It is possible to sample digesta passing from the omasum to the abomasum,[28,36] hence separating the forestomachs from the remainder of the tract, but this point has many associated problems and most research has involved sampling at the anterior duodenum and terminal ileum so as to allow partition of digestive processes into those occurring in the stomach complex, small intestine and large intestine. Two types of techniques are used. Several types of non-absorbable marker substances have been mixed with the feed or given into the rumen, either by twice daily dosing or by continuous infusions. These substances include polyethylene glycol (PEG), ^{51}chromium complex of ethylenediaminetetra-acetic acid (^{51}Cr-EDTA),[26] chromic oxide and ^{103}ruthenium phenanthroline (^{103}Ru-P). The last is largely absorbed onto food particles and so is a marker of the solid phase of the digesta while ^{51}Cr-EDTA and PEG remain in solution and so are markers of the liquid phase. In the digestive tract, the two phases tend to move together except in the rumen from where the liquid phase empties more rapidly than the solid phase. By use of a marker introduced into the rumen, spot samples of digesta taken through a simple intestinal cannula allow calculation of the flow rates of various nutrients to be made from the relative concentration of each nutrient to the marker. A review of techniques is given by Faichney.[29]

The second technique involves the use of re-entrant cannulae in the small intestine, usually near the pylorus and near the ileo-caecal junction.[51] In an experiment, the two parts of a re-entrant cannula are disconnected so that digesta will flow from the proximal end into a container for volume measurement and sampling. The volume of the sample is replaced with similar digesta collected previously and stored and the whole is returned through the distal cannula. The process of collection and the close proximity of experimenters during a 24 h collection period usually depresses flow of digesta through the tract for 24 h or more and so either this technique is used together with a non-absorbable marker for correction of the flow rate,[45] or else collections are made for three or more days to allow flow rates to return to normal. Manual collection is very time consuming and several successful automatic collection and sampling devices have been designed[6] (see [51]).

Re-entrant cannulae permit exchange experiments, useful where metabolites labelled with radioactive markers are used, whereby digesta are exchanged between animals at a specific point of the gut. Two sets of re-entrant cannulae permit temporary isolation of a section of intestine.[68]

D. *Liquid feeding*

An interesting new technique that has considerable potential application for sheep research is that of total liquid feeding. Parenteral nutrition has been done in many monogastric animals, but the presence of the rumen in the sheep creates problems which have only recently been solved.[65] A casein solution is infused through an abomasal cannula, and through a rumen cannula are infused separately a solution of volatile fatty acids and a buffer solution with a similar composition to saliva. In addition, a mixture of vitamins and minerals is given twice daily through the abomasal cannula.

The key to successful infusion is control of the rumen pH which is a process normally done by saliva. However, liquid-fed sheep do not salivate to a sufficient extent to buffer the fatty acids and hence infusion of the buffer solution is necessary. Presence of mild acid in the mouth is a stimulus to salivary secretion and the published method might be modified by (a) continuously pumping a small amount of the clear rumen fluid past a pH electrode for monitoring the rumen pH, (b) returning this fluid to the mouth and (c) adding the buffer automatically as indicated by the rumen pH.

Sheep require about 12 days to become adjusted to the infusion procedure. After this, they eat little when food is offered and do not ruminate. A range of mixtures of volatile fatty acids can be infused and they can be given at up to twice the maintenance needs for several months. The rumen is kept from collapsing by the introduction through the cannula of several "pot scourers" which by their slightly abrasive action maintain the papillae on the rumen wall in a relatively normal state. After a period of liquid feeding, sheep can be re-introduced to a normal ration after about a week, provided that the rumen is innoculated with normal rumen fluid.

7.7 Other Techniques

Some other sampling and recording techniques used for sheep are given below.

A. *Electroencephalogram*

Although EEG recordings can be made from electrodes attached to the skin, it is more usual to insert bone screws surgically into the skull and seal them with dental or bone acrylic cement.[61]

B. *Bone biopsies*

A trephine has been used with local anaesthesia on the mandible.[54]

C. *Radiotelemetry*

Several telemetry systems have been used including one which can be used for ECG, EEG, body temperatures, blood pressure, respiration rate, heart rate and rumen contractions.[73]

D. *Temperature*

Rectal temperatures only roughly follow changes in temperature of the rumen and the vascular compartment and hence should be used with caution.[56] Cooper *et al.*[17] suggested that vaginal temperatures (>10 cm) are more reliable.

E. *Remote infusions and sampling*

Sheep are sufficiently large that they can carry pumps weighing one or more kg. Several pumps have been used for blood and rumen fluid.[18,31,46] They can sample jugular blood without the need for systemic anticoagulation. Another pump system samples rumen fluid but is not portable.[14]

F. *Sensory deprivation*

Some behavioural experiments involve sensory deprivation. Surgical techniques have been used to remove the senses of smell, taste, and touch in the lips and muzzle.[5] Totally blind sheep are grossly abnormal in behaviour and it is preferable to use blinkers to impair the sense of sight.[5] Short-term impairment of smell can be achieved by irrigating the nasal mucosa with 1% zinc sulphate and 3% procaine.[70]

G. *Wool growth*

This parameter is of considerable importance economically and there are three recognized methods for its measurement.[95] In one, a patch 10 cm^2 is tattooed with Indian ink on the midflank side of the sheep. Then at intervals of at least 7 days the area is clipped with fine electric clippers and the wool is collected and weighed. The second method consists of giving doses of ^{35}S-cystine, an amino acid which comprises 9 to 13% of the wool fibre, at intervals of at least 3 days and preferably at least 7 days.[25] After allowing at least 10 days after the last dose for the last labelled zone to pass out of the follicle, a sample of wool is taken, the labelled zones are detected by autoradiography on film and the distance separating the zones is measured. The third method which is suited more to the Merino than to the British breed sheep is dye banding.[16] It consists of applying a solution of black dye (Durafur Black R) with a Pasteur pipette to wool fibres at skin level along a line. The dye stains the wool and penetrates the follicle to a depth of only 50 μ. Applications can be made at intervals of not less than 3 weeks.

H. *Semen*

Rams can be taught to ejaculate into an artificial vagina while mounting a ewe on oestrus but the most common method for collecting semen is to use electrical stimulation. A bipolar electrode is passed into the rectum and an electric current is passed across the electrodes.[9,13] Massage of the seminal vesicles per rectum with a finger will also cause ejaculation in many rams.[66]

I. *Artificial insemination*

See [78].

J. *Egg transfer*

The optimum time for removing eggs is 4 to 5 days. For culture, homologous serum is preferable to TCM199.[76]

Fertilized eggs have been frozen successfully[91] but there has been little success in *in vitro* fertilization of eggs.[11]

K. *Gnotobiotic lambs*

These have been reared successfully for up to 5 months.[1]

L. *Milk*

Techniques which have been used for estimating rate of production include weighing of lambs before and after suckling, measuring the rate of body water dilution using tritiated water, test hand milking and test machine milking.[23]

M. *Nasal secretions*

Tampons left in for 20–30 min allow collection of 0·5–1·5 ml of fluid when squeezed out.[82]

N. *Lens electrodes*

Scleral contact lenses for recording of the electroretinograph have been used successfully.[94]

References

1. Alexander, T. J. L., Lysons, R. J., Elliott, L. M. and Wellstead, P. D. (1973). Techniques for rearing gnotobiotic lambs. *Lab. Anim.*, **7**: 239–254.
2. Allden, W. G. and Jennings, A. C. (1969). The summer nutrition of immature sheep: The nitrogen excretion of grazing sheep in relation to supplements of available energy and protein in a mediterranean environment. *Aust. J. agric. Res.*, **20**: 125–140.
3. Andrews, E. J. and Hughes, H. C. (1973). Thromboembolic sequalae to indwelling Silastic cannulas in sheep arteries. *J. biomed. mater. Res.*, **7**: 137–144.
4. Arnold, G. W. (1960). Harness for the total collection of faeces from grazing ewe and wether sheep. *Anim. Prod.*, **2**: 169–173.
5. Arnold, G. W. (1966). The special senses in grazing animals. 2. Smell, taste, and touch and dietary habits in sheep. *Aust. J. agric. Res.*, **17**: 531–542.
6. Axford, R. F. E., Evans, R. A. and Offer, N. W. (1971). An automatic device for sampling digesta from the duodenum of the sheep. *Res. vet. Sci.*, **12**: 128–131.

7. Bechet, B. (1978). Enregistrement des activities alimentaires et meryciques des ovins au paturage. *Ann. Zootech.*, **27**: 107–113.
8. Binns, R. M., Licence, S. T., Symons, D. B. A., Gurner, B. W., Coombs, R. R. A. and Walters, D. E. (1979). Comparison of the direct antiglobulin rosetting reaction (DARR) and direct immunofluorescence (DIF) for demonstrations of sIg-bearing lymphocytes in pigs, sheep and cattle. *Immunology*, **36**: 549–555. (See also Binns, R. M. (1978). *J. immunolog. Methods*, **21**: 197–210; and Outteridge, P. M., Fahey, K. J. and Lee, C. S. (1981). *Aust. J. exp. Biol. med. Sci.*, **59**: 143–155.)
9. Blackshaw, A. W. (1954). (See reference 3 in Chapter 5.)
10. Bondioli, K. R. and Wright, R. W. (1980). Influence of culture media on *in vitro* fertilization of ovine tubal oocytes. *J. anim. Sci.*, **51**: 660–667.
11. Borrie, J. and Mitchell, R. M. (1960). The sheep as an experimental animal in surgical science. *Br. J. Surg.*, **47**: 435–445.
12. Budtz-Olsen, O. E., Dakin, H. C. and Morris, R. J. H. (1960). Method for continuous urine collection from unrestrained wethers and other large animals. *Aust. J. agric. Res.*, **11**: 72–74.
13. Cameron, R. D. A. (1977). Semen collection and ejaculation in the ram: The effect of method of stimulation on response to electroejaculation. *Aust. vet. J.*, **53**: 380–383.
14. Canaway, R. J., Terry, R. A. and Tilley, J. M. A. (1965). An automatic sampler of fluids from the rumen of fistulated sheep. *Res. vet. Sci.*, **6**: 416–422.
15. Chambers, A. R. M., White, I. R., Russel, A. J. F. and Milne, J. A. (1976). Instruments for sampling and measuring the volume output of urine from grazing female sheep. *Med. Biol. Eng.*, **14**: 665–670.
16. Chapman, R. E. and Wheeler, J. L. (1963). Dye-banding: A technique for fleece growth studies. *Aust. J. Sci.*, **26**: 53–54.
17. Cooper, K. E. *et al.* (1979). (See reference 92 in Chapter 3.)
18. Corbett, J. L., Farrell, D. J., Leng, R. A., McClymont, G. L. and Young, B. A. (1971). Determination of the energy expenditure of penned and grazing sheep from estimates of carbon dixoide entry rate. *Br. J. Nutr.*, **26**: 277–291.
19. Corbett, J. L., Leng, R. A. and Young, B. A. (1969). Measurement of energy expenditure by grazing sheep and the amount of energy supplied by volatile fatty acids produced in the rumen. In *Energy Metabolism of Farm Animals* (Ed. by K. L. Blaxter, J. Kielanowski and G. Thorbeck), pp. 177–185. Oriel Press; Newcastle-upon-Tyne.
20. Corbett, J. L., Lynch, J. J., Nicol, G. R. and Beeston, J. W. U. (1976). A versatile peristaltic pump designed for grazing lambs. *Lab. Pract.*, **25**: 458–462.
21. Denton, D. A. (1957). The effect of variations in blood supply on the secretion rate and composition of parotid saliva in Na^+-depleted sheep. *J. Physiol.*, **135**: 227–244.
22. Dick, A. T. and Mules, M. W. (1954). (See reference 6 in Chapter 6.)
23. Doney, J. M., Peart, J. N., Smith, W. F. and Louda, F. (1979). A

consideration of the techniques for estimation of milk yield by suckled sheep and a comparison of estimates obtained by two methods in relation to the effect of breed, level of production and stage of lactation. *J. agric. Sci.*, **92**: 123–132.
24. Dooley, P. C., Morris, R. J. H., Williams, V. J. and Boffinger, V. J. (1974). An investigation into the precision of micro-haematocrit determinations of sheep blood. *Aust. J. exp. Biol. med. Sci.*, **52**: 663–677.
25. Downes, A. M. and Lyne, A. G. (1961). Studies on the rate of wool growth using [^{35}S]-cystine. *Aust. J. biol. Sci.*, **14**: 120–130.
26. Downes, A. M. and McDonald, I. W. (1964). The chromium-51 complex of ethylenediamine tetraacetic acid as a soluble rumen marker. *Br. J. Nutr.*, **18**: 153–162.
27. Edwards, A. W. T. (1966). Regional pulmonary function by lobar spirometry in unanaesthetized sheep. *J. appl. Physiol.*, **21**: 388–392.
28. Engelhardt, W. V. and Hauffe, R. (1975). Role of the omasum in absorption and secretion of water and electrolytes in sheep and goats, pp. 216–230. (In reference 53.)
29. Faichney, G. J. (1975). The use of markers to partition digestion within the gastro-intestinal tract of ruminants, pp. 277–291. (In reference 53.)
30. Faichney, G. J. and Colebrook, W. F. (1979). A simple technique to establish a self-retaining rumen catheter suitable for long-term infusions. *Res. vet. Sci.*, **26**: 385–386.
31. Farrell, D. J., Corbett, J. L. and Leng, R. A. (1970). Automatic sampling of blood and ruminal fluid of grazing sheep. *Res. vet. Sci.*, **11**: 217–220.
32. Geddes, L. A. (1970). *The Direct and Indirect Measurement of Blood Pressure.* Year Book Medical Publishers; Chicago.
33. Glen, J. B. (1970). Indirect blood pressure measurement in anaesthetized animals. *Vet. Rec.*, **87**: 349–354.
33. Greenwood, B. (1977). Haematology of the sheep and goat, pp. 305–344. In *Comparative Clinical Haematology* (edited by R. K. Archer and L. B. Jeffcott). Blackwell; Oxford.
34. Greenwood, B. and Kerry, P. J. (1979). Prostaglandins and the sources of their production in a mild inflammatory lesion in sheep. *J. Physiol.*, **288**: 379–391.
36. Hecker, J. F. (1974). (See reference 19 in chapter 1.)
37. Hecker, J. F. Unpublished observations.
38. Hecker, J. F. (1979). Thrombus formation and renal infarction induced by aortic catheters in sheep. *Trans Am. Soc. artif. internal Organs*, **25**: 294–297.
39. Hecker, J. F. (1981). Thrombogenicity of tips in umbilical catheters. *Pediatrics*, **67**: 467–471.
40. Hecker, J. F., Fisk, G. C., and Farrell, P. C. (1976). Measurement of thrombus formation on intravascular catheters. *Anaesth. Intens. Care*, **4**: 225–231.
41. Hecker, J. F., Fisk, G. C., Gupta, J. M., Abrahams, N., Cockington, R. A. and Lewis, B. R. (1979). Thrombus formation on catheters in new born lambs. *Anaesth. Intens. Care*, **7**: 239–243.

42. Hoover, W. H., Young, P. J., Sawyer, M. S. and Apgar, W. P. (1970). Ovine physiological responses to elevated ambient carbon dioxide. *J. appl. Physiol.*, **29**: 32–35.
43. Ingleton, J. W. (1971). Faeces collection in young male lambs and wether sheep. *J. Br. Grasslands Soc.*, **26**: 103–106.
44. Itabisashi, T. (1970). Potential changes that accompany the movement of the ruminant stomach, pp. 42–51. (In reference 69.)
45. Kay, R. N. B. and Pfeffer, E. (1970). Movements of water and electrolytes into and from the intestine of the sheep, pp. 390–402. (In reference 69.)
46. Kennaway, D. J.,Porter, K. J. and Seamark, R. F. (1978). Changes in plasma tryptophan and melatonin content in penned sheep. *Aust. J. biol. Sci.*, **31**: 49–52.
47. Kerry, P. J. (1976). The isolation of ovine lymphocytes and granulocytes from whole blood using hydroxyethylcellulose. *Res. vet. Sci.*, **21**: 356–357.
48. Landa, J. F. *et al.* (1975). (See reference 227 in Chapter 3.)
49. Lee, C. S. and Outteridge, P. M. (1976). The identification and ultrastructure of macrophages from the mammary gland of the ewe. *Aust. J. exp. Biol. med. Sci.*, **54**: 43–55.
50. Luedke, A. J. (1965). Procedure for counting ovine leukocytes by electronic means. *Am. J. vet. Res.*, **26**: 1249–1253.
51. MacRae, J. C. (1975). The use of re-entrant cannulae to partition digestive function within the gastro-intestinal tract of ruminants, pp. 261–276. (In reference 53.)
52. MacRae, J. C. and Wilson, S. (1977). The effects of various forms of gastrointestinal cannulation on digestive measurements in sheep. *Br. J. Nutr.*, **38**: 65–71.
53. McDonald, I. W. and Warner, A. C. I. (1975). editors. *Digestion and Metabolism in the Ruminant.* The University of New England Publishing Unit, Armidale, Australia.
54. McDougall, J. G., Coghlan, J. P., Scoggins, B. A. and Wright, R. D. (1974). Effect of sodium depletion on bone sodium and total exchangeable sodium in sheep. *Am. J. vet. Res.*, **35**: 923–929.
55. Masters, D. G. and Moir, R. J. (1980). Provision of zinc to sheep by means of an intraruminal pellet. *Aust. J. exp. Agric. anim. Husb.*, **20**: 547–551.
56. Mendel, V. E. and Raghavan, G. V. (1964). (See reference 267 in Chapter 3.)
57. Michell, A. R. (1977). An inexpensive metabolic harness for female sheep. *Br. vet. J.*, **133**: 483–485.
58. Morag, M. and Robertson Smith, D. (1971). Problems of long-term jugular cannulation in the ewe. *Res. vet. Sci.*, **12**: 192–194.
59. Moseley, G. and Jones, J. R. (1979). A technique for sampling total rumen contents in sheep. *Res. vet. Sci.*, **27**: 97–98.
60. Moule, G. R. (1965). *Field Investigations with Sheep: a Manual of Techniques.* The Commonwealth Scientific and Industrial Research Organization; Melbourne.

61. Mullenax, C. H. and Dougherty, R. W. (1964). Systemic responses of sheep to high concentrate of inhaled carbon dixoide. *Am. J. vet. Res.*, **25**: 424–440.
62. Nichols, G. de la M. (1966). Radiotransmission of sheep's jaw movements. *N.Z. J. agric. Res.*, **9**: 468–473.
63. Nichols, G. de la M. and O'Reilly, E. D. (1966). Transmission and reception of sheep heart-rate. *N.Z. J. agric. Res.*, **9**: 460–467.
64. Olsen, J. D. (1979). Method for repeated or prolonged rumen infusion without establishing an open fistula. *Am. J. vet. Res.*, **40**: 730–732.
65. Orskov, E. R., Grubb, D. A., Wenham, G. and Corrigall, W. (1979). The sustenance of growing and fattening ruminants by intragastric infusion of volatile fatty acid and protein. *Br. J. Nutr.*, **41**: 553–558.
66. Peeters, G. and deBackere, M. (1963). Influence of massage of the seminal vesicles and ampullae and of coitus on water diuresis of the ram. *J. Endocr.*, **26**: 249–258.
67. Perlman, B. B. and Freedman, L. R. (1971). Experimental endocarditis: 2. Staphylococcal infection of the aortic valve following placement of a polyethylene catheter in the left side of the heart. *Yale J. Biol. Med.*, **44**: 206–213.
68. Phillips, W. A., Webb, K. E. and Fontenot, J. P. (1978). Isolation of segments of the jejunum and ileum for absorption studies using double re-entrant cannulae in sheep. *J. anim. Sci.*, **46**: 726–731.
69. Phillipson, A. T. (1970), editor. *Physiology of Digestion and Metabolism in the Ruminant*. Oriel Press; Newcastle-upon-Tyne.
70. Poindron, P. (1974). Methode de suppression reversible de l'odorat chez la Brevis et verification de l'anosmie au moyen d'une epreuve comportementale. *Ann. Biol. anim. Biochem. Biophys.*, **14**: 411–416.
71. Raabe, R. (1968). An efficient method of excreta collection from caged sheep. *Lab. Practice*, **17**: 217–218.
72. Reid, C. S. W. (1963). Diet and the motility of the forestomachs of the sheep. *Proc. N.Z. Soc. anim. Prod.*, **23**: 169–187.
73. Riley, J. L. (1971). A radiotelemetry system for transmitting physiologic data from animals. *Am. J. vet. Res.*. **32**: 155–161.
74. Robinson, S. M., Cadwallader, J. A. and Hill, P. McN. (1978). An animal model for the study of regional lung function. *J. appl. Physiol.*, **45**: 320–324.
75. Romagnoli, A. (1956). Indirect blood pressure measurement in sheep and goats employing the electronic plethysmograph: Validation against the capacitance manometer. *Br. vet. J.*, **112**: 247–252.
76. Rowson, L. E. A. (1971). Egg transfer in domestic animals. *Nature*, **233**: 379–381.
77. Ruckebusch, Y. and Tomov, T. (1973). The sequential contractions of the rumen associated with eructation in sheep. *J. Physiol.*, **235**, 447–458.
78. Salamon, S. (1976). *Artificial Insemination of Sheep*. University of Sydney Press; Sydney. (See also Morrant, A. J. and Dun, R. B. (1960). *Aust. vet. J.*, **36**: 1–7.)

79. Schultz, R. A., Pretorius, P. J. and Terblanche, M. (1972). An electrocardiographic study of normal sheep using a modified technique. *Onderstepoort J. vet. Res.*, **39**: 97–106.
80. Smith, J. B., McIntosh, G. H. and Morris, B. (1970). The traffic of cells through tissues: a study of peripheral lymph in sheep. *J. Anat.*, **107**: 87–100.
81. Smith, P. T. (1978). Electrocardiograms of 32 2-tooth Romney rams. *Res. vet. Sci.*, **24**: 283–286.
82. Smith, W. D. (1975). The nasal secretion and serum antibody response of lambs following vaccination and aerosol challenge with parainfluenza 3 virus. *Res. vet. Sci.*, **19**: 56–62.
83. Spanos, H. G. and Hecker, J. F. (1976). Thrombus formation on indwelling venous cannulae in sheep: Effects of time, size and materials. *Anaesth. Intens. Care*, **4**: 217–224.
84. Spanos, H. G. and Hecker, J. F. (1979). Effects of heparin, asprin and dipyridamole on thrombus formation on venous catheters. *Anaesth. Intens. Care*, **7**: 244–247.
85. Stacy, B. D. and Wilson, B. W. (1970). Acidosis and hypercalciuria: Renal mechanisms affecting calcium, magnesium and sodium excretion in the sheep. *J. Physiol.*, **210**: 549–564.
86. Steel, E. G. (1974). Evaluation of electronic blood platelet counting in sheep and cattle. *Am. J. vet. Res.*, **35**: 1465–1467.
87. Wanner, A. and Reinhart, M. E. (1978). Respiratory mechanics in conscious sheep: Response to methacholine. *J. appl. Physiol.*, **44**: 479–482.
88. Weaver, A. D. (1971). Seasonal variations in ovine urinary constituents, with special reference to pH and potassium concentration. *Am. J. vet. Res.*, **32**: 813–816.
89. Webster, A. J. F. (1967). Continuous measurement of heart rate as an indicator of the energy expenditure of sheep. *Br. J. Nutr.*, **21**: 769–785.
90. Weston, R. H. (1959). The efficiency of wool production of grazing Merino sheep. *Aust. J. agric. Res.*, **10**: 865–885.
91. Willadsen, S. M. (1980). (See reference 399 in Chapter 3.)
92. Winter, H. (1965). Comparison of haematological stains in sheep and man. *Aust. vet. J.*, **41**: 14–16.
93. Winter, H. (1966). Changes of the red blood cell hemogram in post-hemorrhagic anemia in sheep. *Am. J. vet. Res.*, **27**: 891–897.
94. Witzel, D. A., Johnson, J. H., Pitts, D. G. and Smith, E. L. (1976). Scleral contact lens electrodes for electroretinography in domesticated animals. *Am. J. vet. Res.*, **37**: 983–985.
95. Yeates, N. T. M. *et al.* (1975). (See reference 36 in Chapter 2.)
96. Young, B. A. (1966). A simple method for the recording of jaw movement patterns. *J. Inst. Anim. Technicians*, **17**: 20–21.
97. Zehr, J. E., Johnson, J. A. and Moore, W. W. (1969). Left atrial pressure, plasma osmolality and ADH levels in the unanaesthetized ewe. *Am. J. Physiol.*, **217**: 1672–1680.

Chapter 8

Diseases

All animals are subject to a variety of diseases and sheep are no exception. The majority of these diseases are uncommon and usually only a few, such as internal parasites, are of continual concern to the farmer. Under laboratory conditions, sheep usually are healthier than when on farms for the reason that most of the diseases which can affect sheep on farms are unlikely to occur in the laboratory.

In this chapter, some sheep diseases are discussed. The selection has been made on a basis primarily of the diseases which are known to occur in the laboratory, which have a reasonable likelihood of occurring there, which are likely to be present in sheep purchased for research purposes, or which may otherwise be of interest to research workers. Veterinary advice should be sought if there is a suspicion that diseases are causing problems in laboratory sheep.

8.1 Infectious diseases[7,26]

Sheep faeces contain spores of several clostridial organisms including those of *Clostridium tetani*, *Cl. novyi*, *Cl. chauvoei* and *Cl. perfringens* and it is routine management on farms in many areas for lambs to be immunized with a multivalent vaccine against several of these organisms. Losses of sheep from clostridial diseases (tetanus, gas gangrene, infectious necrotic hepatitis (black disease) and enterotoxaemia (pulpy kidney) do occur on farms and occasionally losses in individual outbreaks can be high. Often the outbreak can be linked to poor hygiene, infestations with internal parasites or introduction of animals to lush green crops. Under laboratory conditions, hygiene should be good, infestation with internal parasites is negligible and lush green crops are not fed and so these clostridial diseases are unlikely to occur. There is no report of these diseases in laboratory sheep and so the need for vaccinating laboratory sheep against them is questionable.

Forms of pneumonia can be a problem. There are various causes. Pneumonia can develop as a result of infestation with lung worms (see below). Pasturella infection of the lung (possibly subsequent to parainfluenza virus or *Chlamydia* species infection) causes enzootic pneumonia (acute exudative pneumonia) in feedlot lambs[7] but with the exception of the possibility of its occurrence in animals moved long distances to the laboratory this pneumonia should not affect laboratory sheep. It responds well to antibiotic treatment.

Two other forms of pneumonia are caused by viruses and are each characterized by long incubation periods and long clinical courses. Progressive interstitial pneumonia (Maedi) occurs in Iceland and northern Europe and may also occur in North America.[7,18] Signs are a slow advance of listlessness, emaciation and dyspnoea but there are no signs of excess pulmonary fluid. The lungs are mottled grayish-pink, firm and large and about 2–3 times the normal weight. Histologically, there is thickening of the alveolar walls and accumulation of mononuclear cells in the interstitium. It is most common in older sheep. The other viral pneumonia is pulmonary adenomatosis (*Jaagsiekte*) and is found in South Africa, Europe and Asia. Lungs are similar in appearance to those with progressive interstitial pneumonia but histologically there are characteristic adenomatous growths of alveolar epithelium into the alveolar spaces. It is believed to be a viral-induced cancer.[37]

The other respiratory problem is that of chronic lung abscesses. These are more common in older sheep and are resultant from haematological spread of infection at other sites, from lung worms or from the occasional migration of parasites such as liver fluke to the lungs. These abscesses can cause chronic ill health or they may become walled off and calcified after a time.

Sheep, especially Merinos, may suffer from the disease of Caseous Lymphadenitis (cheesy glands). This is an infection of lymph glands with the lowly pathogenic organism *Corynebacterium pseudotuberculosis* (*Cor. ovis*). It has an unusual mode of transmission. The nodes first affected are the peripheral lymph nodes (usually the prescapular or prefemoral nodes) which gradually fill with pus. A layer of concentric tissue is laid down around the pus but this becomes necrotic causing another layer to be laid down which in turn becomes necrotic. Hence on sectioning an affected node there are concentric layers of inspisated greenish pus. An affected node can reach a size of 5 cm. When a sheep is shorn, an affected node is liable to be incised by the shears which then infect other sheep through shearing cuts. The disease is not found in sheep which have never been shorn and its incidence increases with the age of the sheep (i.e. with the number of shearings). It has little effect on the health of sheep unless (rarely) it becomes a generalized infection. If an enlarged superficial lymph node is found in a laboratory sheep, surgical excision of affected tissues could be performed.

Infection with the blood parasite *Eperythrozoon ovis* is common in sheep, but is not often recognized due to the difficulty of finding the parasite in blood smears. The reason for this difficulty is that there are cycles of high parasitism but parasites are rare in blood smears by the time that the resultant anaemia is apparent. The disease is spread by biting insects and affected animals cure spontaneously in 2–3 months, although some can remain carriers. Subclinical infection is more common than infection which produces obvious symptoms of anaemia. It responds to tetracycline antibiotics.

Contagious opthalmia (keratoconjunctivitis or "pink eye") is spread by flies in the field but should not occur in the laboratory. It responds to local antibiotics. A white opacity of the cornea is a sign that a sheep has previously suffered from this infection.

Footrot[6] is an epizootic hoof infection which is common in sheep in some areas. It starts as a local dermatitis at the skin-horn junction caused by *Fusiformis necrophorus* which is complicated by invasion with *F. nodosus*. *Fusiformis nodosus* is not free-living in soil and so campaigns are conducted in many areas to eliminate the disease by treatment of all affected sheep. It may be confused with the less serious condition, foot abscess, which is caused by *F. necrophorus* followed by invasion of tissue with *Cor. pyogenes* or *Escherichia coli*. If acute lameness occurs in a laboratory sheep, the affected hoof should be trimmed, any pus drained and antibiotic treatment given parenterally. Local treatment, bathing the hoof in 5% formalin or 5% copper sulphate, may be done.

An aspergillus "epidemic" of abortions following foetal surgery, possibly spread by mouldy feed, has been reported.[48] Abortion of ewes due to disease is usually rare but it may be caused by *Vibrio foetus*, a chlamydial organism (enzootic abortion of ewes[29]) or occasionally by *Brucella ovis*. Although *Br. ovis* has only very low pathogenicity in ewes, it is more pathogenic in rams where it causes epididymitis and infertility. The epididymitis is usually unilateral. Although there is a complement fixation test for ovine brucellosis, diagnosis of suspected brucellosis in rams is best done by careful palpation of the testes, comparing one with the other.

Two other viral diseases which deserve mention are contagious ecthyma and scrapie. Contagious ecthyma (contagious pustular dermatitis or "orf") affects the skin just above the hoof causing lameness. It is possible that it could occur in the laboratory although spread is usually by contact with affected animals or contaminated objects. Scrapie is an unusual viral disease which has been much studied as it has similarities with multiple sclerosis. It is caused by a virus which withstands boiling for 30 min and soaking in 20% formalin for 6 months. It is the smallest infective particle known and can be passaged to mice. In sheep, the incubation period is up to 48 months and symptoms are an incessant itching, mental depression, muscular tremors and locomotion incoordination. Mortality is 100%. In the United States and Great Britain, a slaughter programme has caused the disease to almost disappear.

Few sheep diseases are zoonoses. Ovine brucellosis has not been reported to affect humans. Man can develop a mild skin infection from contagious ecthyma. The only disease in laboratory sheep which has been reported to have infected humans is Q fever (caused by *Coxiella burnetti*) which was endemic in sheep in Wisconsin, USA where it produced upper respiratory infections and abortions.[17]

8.2 Internal Parasites

A. *Cestodes*

Several types of adult tape worms are found in sheep intestines and, although they can be spectacular (*Monezia expansa* can reach a length of 7 m), they cause little trouble except in young lambs at pasture.

Cysts of several tapeworms may be found during surgery or at autopsy. A large cyst with a scolex (head) protruding into the cyst is likely to be that of *Taenia hydatigena*. It is usually found in the peritoneal cavity where it does no harm. Masses of small cysts contained in a thin capsule and containing granular material are hydatid cysts from the dog tape worm *Echinococcus granulosus*. They can be found in various organs including the liver and lungs. They may die and become calcified or may continue to propagate by budding. They cause problems when large. Their incidence varies depending on public health measures taken to control this worm in dogs. The cysts in sheep are harmless to humans. To break the life cycle, raw sheep offal should never be fed to dogs.

Intermediate stages of two other dog tape worms reside in skeletal and cardiac muscles. *Cystercercus ovis* infection is from the worm *Taenia ovis* and cysts are most common in the diaphragm and masseter muscles. With *Carcosporidiosis*, small cysts are the intermediate stage of the worm *Sarcocystis tenella*. Eventually these cysts die and become calcified.

B. *Nematodes*

Nematode infections are more serious than cestode infections. *Haemonchus contortus*, called the "barbers pole worm" because of the white uterus which spirals along the red body of the female, lives in the abomasum. The females (20–30 mm) are longer than the males (10–20 mm). *Ostertagia* species (8–12 mm) and *Trichostrongulus* species (males 3–5 mm, females 4–6 mm) are the other main nematodes in the abomasum. In the small intestine are found *Trichostrongulus colubriformis* (4–8 mm), *Nematodirus* species (15–20 mm) and *Cooperia* species. The smaller worms can be found in very large numbers, but being fine and colourless are difficult to see without magnification. *Oesophagostomum columbianum* (the "nodule worm") is larger, lives in the colon, and has an intermediate stage which burrows into the wall of the colon. After infestation for a few months with this worm, sheep develop an allergic reaction causing death of the parasite and formation of a pus nodule which gradually becomes calcified. Older sheep may have many calcified nodules in the colon but the only significance is that they make the gut wall unsuitable for the manufacture of surgical gut and sausage casings.

The other nematodes of importance are the lung worms of which the largest and also the most important is *Dyctocaulus filaria*. The larvae on ingestion pass through the intestinal wall and then go via the blood stream to the lung where they live in the smaller bronchi. They can reach 4 cm in length. "Husk" is the term given to the chronic bronchitis that they produce.

All sheep brought into the laboratory are likely to be infected with worms. Worms are most serious in young sheep in which heavy infections cause loss of weight, loss of appetite, diarrhoea and anaemia. Worms are readily killed by modern anthelmintics such as the phenothiazine and thiabendazole derivatives, although resistance to some anthelmintics is known. Many anthelmintics are ineffective against lung worms but diethylcarbamazine, mebendaloze, fenbendazole and levamisole are effective against both lung and most intestinal worms. Some of these drugs can be injected.

Anthelmintics tend to kill only the mature parasites and this allows immature stages, sometimes prevented from final development by the presence of adult worms, to mature. Hence when sheep are first introduced into the laboratory, they should be dosed twice at an interval of about 2 weeks. As the first stage of most worm species has to climb onto grass from where it can be ingested, the chances of reinfection of sheep in the laboratory are small.

The level of nematode infestation can be determined by mascerating about 3 g of faecal pellets in 50–100 ml of a solution with high specific gravity (saturated salt or sucrose solution). Almost all nematode eggs have a lower specific gravity than plant particles and so the eggs will float to the surface if the mascerated solution is placed in a tall collection cylinder and left to stand. For counting, eggs are collected with a pipette and placed in a counting chamber.

C. *Trematodes*

The liver fluke (*Fasciola hepatica*) causes acute or chronic infection. This fluke has as its intermediate host several types of freshwater snails which live in wet or swampy land. Infestations can be heavy and can cause marked haemorrhage and necrosis in young sheep while the fluke are passing through the liver. This in itself can cause death or it can predispose to rapid death from "black disease" (p. 196). Chronic fascioliasis when severe results in anaemia and if the level of plasma albumin is sufficiently low, oedema is seen, primarily under the jaw. A slight infection results in fibrosis of the bile ducts.

The key to fascioliasis is the snail intermediate host and most farmers will fence off parts of fields where snails are common so as to prevent sheep losses. It is unlikely that sheep suffering from a severe infection with fluke will be found in the laboratory but it is possible that laboratory sheep may have a few fluke.

8.3 External Parasites

Lice and ked are of concern to farmers as they cause sheep to itch and rub for long periods against fences, trees and other objects. The wool becomes matted and much of it is lost from the sides of animals. There are two types of lice, the biting louse (*Damilinia ovis*) and the sucking louse (*Linognathus pedalis*) which are virtually colourless, only about 2 mm long and hence difficult to see. The sheep ked (*Melophagus ovinus*) is a wingless fly and is longer (4–7 mm). Diagnosis of infestation is by finding lice or ked but this can be difficult if the wool is long and much searching of parted wool may be needed.

Farmers control lice and ked by spraying or dipping of animals soon after shearing, as at other times the wool is too dense to allow effective penetration of insecticides. If a heavy infestation is found in laboratory sheep, shearing followed by spraying with an insecticide should be done. Most well known insecticides are suitable. If sheep are kept in an animal house, insecticide should persist in the wool for a considerable time as there would be little leaching by rain.

An unusual infestation with an external parasite is in the nasal sinuses with "bots" which are the larval stage of the fly *Oestrus ovis*. Eggs are laid by the female fly on the external nares and the larvae on hatching move to the sinuses where they live for from 1–10 months. They then crawl out, drop to the ground and pupate. They cause irritation, nasal discharge and sneezing. Their presence in laboratory sheep often is not recognized until an animal is anaesthetized with an inhalational anaesthetic when they crawl from the nares. Treatment is rarely attempted and it is unlikely that fresh infestation would occur under laboratory conditions.

Blow fly strike (*cutaneous myasis*) is a serious condition of field sheep and can affect laboratory sheep fitted with intestinal cannulae which extrude moisture. It is caused by certain species of blow flies being attracted to and laying eggs on moist areas of wool. The eggs hatch into maggots which burrow into the wool against the skin where they cause an intense irritation and a moist exudatc which attracts more flies to lay eggs. Thus the lesion enlarges. Affected sheep spend much time trying to bite or rub the affected parts. They tend not to eat and hence lose condition rapidly.

If fly strike is found in a sheep, the affected wool should be removed together with a strip of wool about 4 cm wide around the affected area. Then an insecticide should be applied to the affected parts and surrounding wool to kill the maggots that are present and those which subsequently hatch. The sheep should then be inspected daily for the next 3–4 days to ensure a complete kill. Prevention involves periodic clipping of wool from around areas which are likely to become moist. These include around the perineum of ewes, the prepuce of rams and wethers and intestinal cannulae. As laboratory sheep rarely become wet, some of the other forms of strike such as body strike are unlikely to occur. Hence jetting (high pressure spraying) of sheep as is done on farms is not required.

8.4 Tumours

Tumours are relatively rare in sheep for several reasons. Most sheep reared for slaughter are killed at a young age. Even breeding animals that are culled and slaughtered for meat are still at a young age (5–8 years) compared with the potential life span of the sheep (15–20 years). Also very few sheep that die on farms are submitted for autopsy. Data from two surveys showing the relative incidences of different types of tumours are given in Table 41.

The most common sheep tumours are leukaemia, pulmonary carcinoma and hepatoma. In the United States and Great Britain, ovine lymphosarcoma is one of the most common causes of condemnations of carcases in abattoirs.[8]

The incidences of some of the sheep tumours are exceedingly variable and this is suggstive of involvement of either environmental carcinogens or else viruses. At least three are caused by viruses.[33] A *c*-type oncornovirus which causes leukaemia in sheep has been isolated and shown to be closely related if not identical to a bovine leukaemia virus. This bovine virus causes leukaemia in sheep.[35] Pulmonary carcinoma, sometimes called *Jaagseikte*, is episodic in South Africa and

Table 41 Incidence of tumours in sheep in two surveys.

System	Tumour	Frequency 14	1
Skin	Melanoma		1
	Papilloma	4	
	Squamous cell carcinoma	3	2
Musculoskeletal	Chondroma		3
	Rhabdomyoma	3	
Blood and lymph	Lymphosarcoma	40	44
Respiratory	Pulmonary adenoma		1
Digestive	Intestinal carcinoma	143	1
	Hepatoma	12	23
	Cholangiocarcinoma	15	9
	Biliary cystadenoma		1
Urinary	Embryonal nephroma	3	
	Renal adenocarcinoma	1	4
Genital	Seminoma	2	
	Granulosa cell tumour	3	
	Uterine adenomas		2
	Sertoli cell tumour	1	
Endocrine	Adrenal adenoma and pheochromocytoma	1	2
	Thyroid adenoma		1
	Pituitary adenoma	1	
Blood and lymph vessels	Angioma	8	3
Nervous	Neurofibroma	1	
	Astrocytoma	1	
Connective tissue	Fibrosarcoma	4	3
	Myxosarcoma	3	
	Osteogenic sarcoma	2	
	Lipoma	1	1
	Synovioma	1	

Iceland and in the latter country was traced to the introduction of a ram from Germany in 1933. This tumour is transmissable and virus particles have been found.[37] Intestinal adenocarcinoma, also a viral tumour, has an uneven distribution and is most common in Iceland and New Zealand.[43]

The only tumour that has been experimentally induced is a sarcoma caused by injecting foetal kidney cells grown in tissue culture and transformed with feline sarcoma virus. It grew well for a time in several lambs and exceedingly rapidly in one lamb that had been thymectomized *in utero*.[47] In most lambs, immunological rejection occurred after a few weeks.

Exposure to high levels of sunlight as is common in Australia causes a squamous cell carcinoma on exposed regions of skin, usually on the ears but sometimes on the nose or perineum.[28] Farmers often remove affected ears by cutting them off and hence its incidence is unknown but it could be 0·2%.[30] Sera from sheep with squamous cell carcinoma contain a factor which nonspecifically inhibits blastogenesis of T lymphocytes, possibly through a B lymphocyte suppression pathway.[27]

Genetics has been incriminated in an interesting sheep tumour. Squamous cell carcinoma is rare on woolly areas of skin but in three related inbred flocks of Australian Merino sheep, there was a high incidence (up to 10% of new cases each year) of this tumour. The suggestion was made that penetrating grass seeds carried epithelial cells into the dermis where in these particular sheep they formed implantation dermoids which later became cancerous.[10]

8.5 Metabolic and Mineral Deficiency Diseases

Several such diseases are of importance. Pregnancy toxaemia (ketosis or twin lamb disease) occurs in heavily pregnant sheep who usually have two or more lambs rather than a single lamb. Ruminants are always in a slightly precarious situation with regard to supplies of glucose (p. 79) and the demands of the near-term foetus for glucose can make the situation critical if the food intake is suddenly limited.[4] Symptoms are inappetence and lethargy leading to coma. Ketone bodies are present in large amounts in blood and urine. The most satisfactory treatment for a ewe with pregnancy toxaemia is to perform a caesarian delivery. Intravenous infusions of glucose are not very effective but may be more efficacious if glucocorticosteroids are also given. Early treatment is more likely to be successful. For prevention, the food intake of ewes should gradually be increased during pregnancy and sudden reductions in weight or quality of food should be avoided. Although fasting of ewes prior to intrauterine surgery might be expected to induce pregnancy toxaemia, from no laboratory has its occurrence been reported. Even so, fasting of ewes prior to intrauterine surgery should be confined to an overnight period.

Sudden introduction of high levels of grain into the diet of ruminants can induce lactic acidosis. Grain contains large amounts of readily fermentable starch which in the rumen is first broken down to lactic acid. In an animal accustomed to grain in the diet, there are large numbers of organisms present in the rumen which break down lactic acid to propionic acid. If these organisms are not numerous, the lactic acid accumulates and the rumen pH drops to a level where much of the rumen microflora is killed and rumen motility is inhibited. Much lactic acid is absorbed and induces laminitis (swelling of the tissues within the hoof). A mild case of lactic acidosis is evidenced by inappetence and lameness and will cure spontaneously but there is no effective treatment for a severe case. To avoid the condition, grain should be introduced gradually into a ration over a period of a week.

Lactating ewes on low calcium diets may suffer from hypocalcaemia (milk fever). Classical signs are a slight hyperexcitability, muscle tremors and a stilted gait, soon followed by dullness, sternal recumbency with the legs under the body or stretched out behind, then coma and death within 6–12 h. Treatment with calcium borogluconate produces a rapid response.

Experimental hypomagnesaemia can be induced in wethers by an intra-abomasal infusion of milk with a low magnesium content.[5] The disease occurs naturally on farms when lush grass pasture is grazed. It is often combined with hypocalcaemia and the signs may not be clear cut.

Selenium deficiency is common in areas with a low selenium content in the soil. It affects lambs from birth to 3 months but usually at 2 to 4 weeks of age. It is known as white muscle disease and takes two forms which affect muscle. When skeletal muscles are predominantly affected, there is unsteadiness and weakness while for cardiac muscle involvement the pulse is rapid and weak and there is pulmonary oedema. The red cell enzyme, glutathione peroxidase, is a selenium dependent enzyme and an indication of the selenium status may be provided by the activity of this enzyme.[2] Treatment is by dosing with vitamin E or preferably with selenium and this is done most conveniently on farms combined with anthelmintic treatment three to four times each year.[50]

Copper deficiency also affects lambs up to 3 months of age causing "enzootic ataxia" in which there is a progressive incoordination of the hind legs due to a disruption of neurone and myelin development in the central nervous system.[44] In copper deficient areas or for copper deficient foods, prevention is provided by an intramuscular injection of copper glycenate to ewes once per year.

Sheep are the most sensitive of all domestic animals to the effects of excess copper. The reason appears to be a lack of homeostatic control over hepatic copper levels as liver copper is closely related to dietary copper levels.[13] The condition occurs in two phases. In the accumulation phase, copper accumulates in the liver and kidneys. There are no specific signs but the plasma levels of liver specific enzymes (especially aspartate amino transferase[45]) are increased and BSP clearance by the liver is decreased. Then there are one or more haemolytic crises with jaundice, haemoglobinuria and focal necrosis of the liver. Kidneys are swollen, black or brown with little demarcation between the cortex and medulla. The trigger for the haemolytic crises is unknown but it may be associated with a build up of isoantibodies.[51]

8.6 Congenital Diseases and Conditions[41]

Several studies have been made of the causes of death of new-born lambs. Some of these deaths are due to factors such as inclement weather or mismothering but others have more basic causes. A large proportion of dead lambs have suffered from haemorrhage into the central nervous system with blood stained cerebrospinal fluid being very common in periparturient deaths and epidural haemorrhage in about 50% of postparturient deaths.[24] Intraventricular haemorrhage which is an important cause of mortality in new-born babies is rare in lambs. In two surveys in which over 8000 dead lambs were examined,[19,25] about 10% had congenital defects which included a high proportion of craniofacial and cardiac defects. Some of these may have been due to the ingestion of poisonous plants but the only such plant that has been identified is *Veratrum californicum* which induces this malformation on day 14 of gestation in about 1% of ewes at pasture and a higher proportion when fed in the laboratory[34] (p. 65).

Two other factors causing foetal malformations are viral infections and heat. Infection with bluetongue disease virus causes hydrancephaly, porencephaly and retinal displasia,[36] while maternal hyperthermia during the last two thirds of pregnancy produces brain cavitation and microencephaly.[9,23] The hyperthermia effect might be due to intra-uterine malnutrition caused by a poor transplacental transfer of nutrients.[12]

Several congenital conditions inherited as recessive lethal or sublethal genes have been studies for their relation to human disease (Table 42). All are rare.

Table 42 Some lethal and sublethal conditions in sheep.

Condition	Breed	Country
Hyperbilinrubinaemia	Southdown	New Zealand, USA[16,21]
Dubin-Johnson syndrome		USA[15]
Generalized glycogen storage disease	Corriedale	New Zealand[32]
Congenital goitre	Merino	Australia[20]
Muscular dystrophy	Merino	Australia[31]
Hereditary "daft lamb" disease (inherited cerebellar cortical atrophy)	Several	Great Britain[46]
Mesangiocapillary glomerulonephritis	Finnish Landrace[3]	
Collagen dysplasia		Norway[22]
Globoid cell leucodystrophy	Merino	Australia[38]

8.7 Toxicology

Sheep have been used for testing the toxicity of several heavy metals and organic substances including cadmium,[52] mercury[39] and lead.[40,42] They have advantages as they can be used to detect environmental contamination of land around industrial complexes from their intake of forage. The wool also is a good adsorbent of atmospheric lead contamination.[49] Unlike man, chronic lead ingestion has no anaemic effect.[11] Many plants in various parts of the world also have toxic effects but they are not included as they tend to be of only local interest.

References

1. Anderson, L. J., Sandison, A. T. and Jarrett, W. F. H. (1969). A British abbattoir survey of tumours in cattle, sheep and pigs. *Vet. Rec.*, **84**: 547–551.

2. Anderson, P. H., Berrett, S. and Patterson, D. S. P. (1979). The biological selenium status of livestock in Britain as indicated by sheep erythrocyte glutathione peroxidase activity. *Vet. Rec.*, **104**: 235–238. (See also Peter, D. W. (1980). *Vet. Rec.*, **107**: 193–196.)
3. Angus, K. W., Sykes, A. R., Gardiner, A. C. and Morgan, K. T. (1974). Mesangiocapillary glomerulonephritis in lambs: 1. Clinical and biochemical findings in a Finnish Landrace flock. *J. comp. Path.*, **84**: 309–317.
4. Baird, G. D. (1977). Aspects of ruminant intermediary metabolism in relation to ketosis. *Biochem. Soc. Trans.*, **5**: 819–927.
5. Baker, R. M., Boston, R. C., Boyes, T. E. and Leaver, D. D. (1979). Variations in the response of sheep to experimental magnesium deficiency. *Res. vet. Sci.*, **26**: 129–133.
6. Barber, D. M. L. (1979). Foot rot in sheep. *Vet. Rec.*, **105**: 194–195.
7. Blood, D. C., Henderson, J. A. and Radostits, O. M. (1979). *Veterinary Pathology*, 5th edition. Bailliere Tindall; London.
8. Bostock, D. E. and Owen, L. N. (1973). Porcine and ovine lymphosarcoma: A review. *J. Nat. Cancer Instit.*, **50**: 933–939.
9. Brown, D. E., Harrison, P. C., Hinds, F. C., Lewis, J. A. and Wallace, M. H. (1977). Heat stress effects on fetal development during late gestation in the ewe. *J. anim. Sci.*, **44**: 442–446.
10. Carne, H. R., Lloyd, L. C. and Carter, H. B. (1963). Squamous carcinoma associated with cysts of the skin in Merino sheep. *J. Path. Bact.*, **86**: 305–315.
11. Carson, T. L., Gelder, G. A. van, Buck, W. B., Hoffman, L. J., Mick, D. L. and Long, K. R. (1973). Effects of low level lead ingestion in sheep. *Clin. Toxicol.*, **6**: 389–403.
12. Cartwright, G. A. and Thwaites, C. J. (1976). (See reference 69 in Chapter 3.)
13. Corbett, W. S., Saylor, W. W., Long, T. A. and Leach, R. M. (1978). Intracellular distribution of hepatic copper in normal and copper-loaded sheep. *J. anim. Sci.*, **47**, 1174–1179.
14. Cordes, D. O. and Shortridge, E. H. (1971). Neoplasms of sheep: A survey of 256 cases recorded at Ruakura animal health laboratory. *N.Z. vet. J.*, **19**: 55–64.
15. Cornelius, C. E., Arias, I. M. and Osburn, B. I. (1965). Hepatic pigmentation with photosensitivity: A syndrome in Corriedale sheep resembling Dubin-Johnson syndrome in man. *Am. J. vet. med. Assn.*, **146**: 709–713.
16. Cornelius, C. E. and Gronwall, R. R. (1968). (See reference 93 in Chapter 3.)
17. Curet, L. B. and Paust, J. C. (1972). Transmission of Q fever from experimental sheep to laboratory personnel. *Am. J. Obstet. Gynecol.*, **114**: 566–568.
18. Cutlip, R. C., Jackson, T. A. and Lehmkuhl, H. D. (1979). Lesions of ovine progressive pneumonia: Interstitial pneumonitis and encephalitis. *Am. J. vet. Res.*, **40**: 1370–1374.

19. Dennis, S. M. and Leipold, H. W. (1979). Ovine congenital defects. *Vet. Bull.*, **49**: 233–239.
20. Dolling, C. E. and Good, B. F. (1976). (See reference 111 in Chapter 3.)
21. Filippich, L. J., English, P. B. and Seawright, A. A. (1977). Comparison of renal function in a congenital hyperbilirubinaemic Southdown sheep and normal sheep. *Res. vet. Sci.*, **23**: 204–212.
22. Fjolstad, M. and Helle, O. (1974). A hereditary dysplasia of collagen tissues in sheep. *J. Path.*, **112**: 183–188.
23. Hartley, W. J., Alexander, G. and Edwards, M. J. (1974). Brain cavitation and microencephaly in lambs exposed to prenatal hyperthermia. *Teratology*, **9**: 299–304.
24. Haughey, K. G. (1973). Vascular abnormalities in the central nervous system associated with perinatal lamb mortality: 1. Pathology. *Aust. vet. J.*, **49**: 1–8.
25. Hughes, K. L., Haughey, K. G. and Hartley, W. J. (1972). Spontaneous congenital developmental abnormalities observed at necropsy in a large survey of newly born dead lambs. *Teratology*, **5**: 5–10.
26. Jensen, R. (1974). *Diseases of Sheep*. Lea and Febiger; Philadelphia.
27. Jun, M. H. and Johnson, R. H. (1979). Suppression of blastogenic response of peripheral lymphocytes by serum from ovine squamous cell carcinoma bearing sheep. *Res. vet. Sci.*, **27**: 161–166.
28. Ladds, P. W. and Entwistle, K. W. (1977). Observations on squamous cell carcinomas of sheep in Queensland Australia. *Br. J. Cancer*, **35**: 110–114.
29. Linklater, K. A. and Dyson, D. A. (1979). Field studies on enzootic abortion of ewes in south east Scotland. *Vet. Rec.*, **105**: 387–389.
30. Lloyd, L. C. (1961). Epithelial tumours of the skin of sheep: Tumours of areas exposed to solar radiation. *Br. J. Cancer*, **15**: 780–789.
31. McGavin, M. D. and Baynes, I. D. (1969). A congenital progressive ovine muscular dystrophy. *Path. Vet.*, **6**: 513–524.
32. Manktelow, B. W. and Hartley, W. J. (1975). Generalized glycogen storage diseases in sheep. *J. comp. Path.*, **85**: 139–145.
33. Moulton, J., editor. (1978). *Tumours in Domestic Animals*, 2nd edition. University of California Press; Berkley, Los Angeles.
34. Mulvihill, J. J. (1972). Congenital and genetic disease in domestic animals. *Science*, **176**: 132–137.
35. Ogura, H., Paulsen, J. and Bauer, H. (1977). Cross-neutralization of ovine and bovine C-type leukaemia virus-induced syncytia formation. *Cancer Res.*, **37**: 1486–1489.
36. Osburn, B. I. and Silverstein, A. M. (1972). Animal model: Bluetongue-vaccine-virus infection in fetal lambs. *Am. J. Path.*, **67**: 211–214.
37. Perk, K., Hod, I., Presentey, B. and Nobel, T. A. (1971). Lung carcinoma of sheep (Jaagsiekte): 2. Histogenesis of the tumor. *J. Nat. Cancer Inst.*, **47**: 197–205.
38. Prichard, D. H., Napthine, D. V. and Sinclair, A. J. (1980). Globoid cell leucodystrophy in Polled Dorest sheep. *Vet. Path.*, **17**: 399–405.

39. Robinson, M. and Trafford, J. (1977). A study of early urinary enzyme changes in mercuric chloride nephropathy in sheep. *J. comp. Path.*, **87**: 275–280.
40. Rolton, C. E., Horton, B. J. and Pass, D. A. (1978). Evaluation of tests for the diagnosis of lead exposure in sheep. *Aust. vet. J.*, **54**: 393–397.
41. Saperstein, G., Leipold, H. W. and Dennis, S. M. (1975). Congenital defects of sheep. *J. Am. vet. med. Assn.*, **167**: 314–322.
42. Sharma, R. M. and Buck, W. B. (1976). Effects of chronic lead exposure on pregnant sheep and their progeny. *Vet. Toxicol.*, **18**: 186–188.
43. Simpson, B. H. (1972). An epidemiological study of carcinoma of the small intestine in New Zealand sheep. *N.Z. vet. J.*, **20**: 91–97.
44. Smith, R. M., Fraser, F. J., Russell, G. R. and Robertson, J. S. (1977). Enzootic ataxia in lambs: Appearance of lesions in the spinal cord during foetal development. *J. comp. Path.*, **87**: 119–128.
45. Suttle, N. F. (1977). Reducing the potential copper toxicity of concentrates to sheep by the use of molybdenum and sulphur supplements. *Anim. Feed Sci. Technol.*, **2**: 235–246.
46. Terlecki, C., Richardson, S., Bradley, R., Buntain, D., Young, G. B. and Pampiglione, G. (1978). A congenital disease of lambs clinically similar to "inherited cerebellar cortical atrophy" (daft lamb disease). *Br. vet. J.*, **134**: 299–307.
47. Theilen, G., Hall, J. G., Prendry, A., Glover, D. J. and Reeves, B. R. (1974). Tumours induced in sheep by injecting cells transformed *in vitro* with feline sarcoma virus. *Transplantation*, **17**: 152–155. (See also Theilen, G. H., Pedersen, N. C. and Higgins, J. (1979). *J. Nat. Cancer. Instit.*, **63**: 389–397.)
48. Turney, S. Z. and Kennedy, J. H. (1967). Fetal cardiovascular surgery: Experimental pulmonary artery atresia in the fetal lamb. *J. Thorac. Cardiovas. Surg.*, **54**: 761–769.
49. Ward, N. I. and Brooks, R. R. (1979). Lead levels in wool as an indication of lead in blood of sheep exposed to automobile emissions. *Bull. environm. Contm. Toxicol.*, **21**: 403–408.
50. Whanger, P. D., Weswig, P. H., Schmidz, J. A. and Oldfield, J. E. (1978). Effects of various methods of selenium administration on white muscle disease, glutathione peroxidase and plasma enzyme activities in sheep. *J. anim. Sci.*, **47**: 1157–1166.
51. Wilhelmsen, C. L. (1979). An immunohematological study of chronic copper toxicity in sheep. *Cornell Vet.*, **69**: 225–232.
52. Wright, F. C., Palmer, J. S., Riner, J. C., Haufler, M., Miller, J. A. and McBeth, C. A. (1977). Effects of dietary feeding of organocadmium to cattle and sheep. *J. agric. Food Chem.*, **25**, 293–297.

Index

E

F

G

H

I

J

K

L

M